Respiratory Physiology

Edited by Ketevan Nemsadze

Published in London, United Kingdom

Supporting open minds since 2005

Respiratory Physiology
http://dx.doi.org/10.5772/intechopen.83122
Edited by Ketevan Nemsadze

Contributors
Chandrasekhar Krishnamurti, Esmaeil Alibakhshi, Luis Lores Obradors, Raffaele Fiorillo, Ana María Sánchez-Laforga, Amparo Villar Cánovas, Ali Qazvini, Mostafa Ghanei, Ibiwunmi Saliu, Evangelisca Akiomon, François Billaut, Ramon F Rodriguez, Rob Aughey, Ketevan Nemsadze

Notice
Statements and opinions expressed in the chapters are these of the individual contributors and not necessarily those of the editors or publisher. No responsibility is accepted for the accuracy of information contained in the published chapters. The publisher assumes no responsibility for any damage or injury to persons or property arising out of the use of any materials, instructions, methods or ideas contained in the book.

First published in London, United Kingdom, 2020 by IntechOpen
IntechOpen is the global imprint of INTECHOPEN LIMITED, registered in England and Wales, registration number: 11086078, 5 Princes Gate Court, London, SW7 2QJ, United Kingdom
Printed in Croatia

British Library Cataloguing-in-Publication Data
A catalogue record for this book is available from the British Library

Additional hard and PDF copies can be obtained from orders@intechopen.com

Respiratory Physiology
Edited by Ketevan Nemsadze
p. cm.
Print ISBN 978-1-83962-325-7
Online ISBN 978-1-83962-326-4
eBook (PDF) ISBN 978-1-83962-327-1

We are IntechOpen,
the world's leading publisher of
Open Access books
Built by scientists, for scientists

5,000+
Open access books available

126,000+
International authors and editors

145M+
Downloads

Our authors are among the

151
Countries delivered to

Top 1%
most cited scientists

12.2%
Contributors from top 500 universities

Interested in publishing with us?
Contact book.department@intechopen.com

Numbers displayed above are based on latest data collected.
For more information visit www.intechopen.com

Meet the editor

Professor Ketevan Nemsadze is an author of more than 100 articles, two monographs, thirteen recommendation/guidelines, and four teaching modules of the Effective Perinatal Care training package in collaboration with the World Health Organization's Regional Office for Europe. She is chief editor of nine textbooks and co-author and editorial board member of thirty-five others. She is a recipient of two copyright certificates and a patent certificate for inventions made in the field of sepsis management. She was also Advisor-Neonatologist of Ukraine presidential initiative "New Life." Professor Nemsadze is a recipient of numerous awards, among them a certificate from the Agency for Educational Development of the Government of the United States of America (AED/Washington) and a Southeast European Medical Forum (SEMF) Award for contribution to medical science development. Professor Ketevan Nemsadze has played a leadership role in the development and management of child health and development programs. She has long-standing working experience at local and international levels involving elaboration and organization of healthcare policy and strategy and development and implementation of guidelines and protocols. From her position as president (2002–2004) of NGO/UNICEF Regional Network for Children of CEE/CIS and Baltic States, Professor Nemsadze has been able to shape policy and strategy to meet the changing needs of child rights.

Contents

Preface

The respiratory system, as a port of entry, plays an important role in the health and disease of individuals. Understanding respiratory physiology can aid practitioners in diagnosing the cause of respiratory symptoms. Physiological factors should be considered before interpreting any medical condition. Clinical application of physiology knowledge can influence the outcome of respiratory disorders.

This book is for medical students, residents, healthcare workers, and scientists. Understanding of the concepts contained in this book will give confidence to provide effective medical care.

Chapters cover aspects of hyperbaric physiology, exercise physiology, skeletal muscle dysfunction in chronic respiratory disease patients, impact of environmental media on the respiratory system, and cognitive impairment in chronic respiratory patients.

Respiratory Physiology is organized into three sections containing five chapters. Section 1 contains an introductory chapter that discusses the general principles of physiology. Section 2 examines respiratory physiology and exercise. It covers the principles of intermittent sprint work as well as the indicators of sarcopenia in patients with chronic respiratory disease. Section 3 describes the impact of environmental media on the respiratory system.

Sections 2 and 3 provides interesting information about exercise, altitude, and environmental pollution and their impact on health and disease.

As the knowledge of physiology is constantly increasing, this book includes interesting concepts and evidence based on experiences from clinical trials.

I thank all the authors for their contributions in making this interesting manuscript.

I hope readers will enjoy this study of physiology and deepen their knowledge of this field.

Ketevan Nemsadze
David Tvildiani Medical University,
Georgia

Section 1

Introduction

Chapter 1

Introductory Chapter: Respiratory Physiology

Ketevan Nemsadze

1. Introduction: What is physiology?

Physiology is the link between the basic sciences and medicine. It integrates the individual functions of all the body's different cells, tissues, and organs into a functional whole, the human body. Understanding the principles of physiology can help us explain the physical and chemical mechanisms that are responsible for the origin, development, and progression of life. Each type of life is from the simplest virus to the largest tree or the complicated human being. The science of *human physiology* attempts to explain the specific characteristics and mechanisms of the human body that make it a living being. The fact that we remain alive is the result of complex control systems. Ultimately, the functions of cells, tissues, and organs must be coordinated and regulated. All of these functions are the essence of the discipline of physiology [1, 2].

Physiology studies dynamic processes of life from the simplest molecules, organelles, cells, tissues to the complex organs and organ systems. The discipline of physiology has been closely interconnected with medicine. Structure and function are related to each other – as in case of anatomy, histology, structural biology and physiology [3].

Medical physiology deals with how the human body functions. Countless molecules, subcellular organelles, cells, tissues, organs and organ systems work coordinately to maintain the homeostasis of a body. It is essential to take a global view of the human body, considering every level of the organizational unit.

Classification of physiology could be according to the organ systems. For many practicing clinicians, physiology may be the function of an individual organ system, such as the cardiovascular, respiratory, or gastrointestinal system. Others focus on the cellular principles that are common to the function of all organs and tissues. This last field has traditionally been called general physiology, but nowadays the term of cellular and molecular physiology.

As a discipline of physiology evolves and new information emerges, there is a tendency to integrate physiological concepts from the level of DNA and epigenetics to the human body, and everything in between [4].

Physiological genomics is the link between the organ and the gene. The grand organizer—the master that controls the molecules, the cells, and the organs and the way they interact—is the genome with its epigenetic modifications. Physiological genomics (or functional genomics) is a new branch of physiology devoted to the understanding of the roles that genes play in physiology [3].

Some important aspects of physiology remain as fundamentally important today as when the pioneers of physiology discovered them a century or more ago. These early observations were generally phenomenological descriptions that physiologists have since been trying to understand at a mechanistic level. In his lectures on the phenomena of life, Claude Bernard noted in 1878 on the conditions of the constancy of life. Claude Bernard introduced the concept of "milieu intérieur" – internal environment. He stated that animals have two environments: the "milieu extérieur" that physically surrounds the whole organism and the "milieu intérieur," in which the tissues and cells of the organism live. The internal environment surrounds and bathes all the anatomic elements of the tissues, the lymph or the plasma. This internal environment is what we today call the extracellular fluid.

Another theme developed by Bernard was that the "fixité du milieu intérieur" (the constancy of the extracellular fluid) is the condition of "free, independent life." Homeostatic mechanisms—operating through sophisticated feedback control mechanisms—are responsible for maintaining the constancy of the milieu intérieur. Homeostasis is defined as a self-regulating process by which an organism maintains internal stability in a constantly changing external condition. Homeostasis regulates the processes in the body in the way that returns critical systems of the body to a set point that is necessary for the organism to survive [3, 5].

Medicine borrows its physicochemical principles from physiology. Study of physiological system structure and function, as well as pathophysiological alterations, has its foundations in physical and chemical laws and the molecular and cellular makeup of each tissue and organ system. If you know how organs and organ systems function in the healthy person, you will find out which components may be malfunctioning in a patient.

2. The importance of respiration

Life starts with breathing. As it was written in the Bible God "breathed into Adam's nostrils the breath of life" and then used part of Adam's ventilatory apparatus – a rib – to give life to Eve. Hippocrates suggested that the primary purpose of breathing was to cool the heart. In the 18th century the true role of breathing began emerged. By the end of the 18th century, chemists and physiologists studied the chemistry of gases and appreciated that combustion, putrefaction, and respiration all involve chemical reactions that consume O_2 and produce CO_2. Boyle, Henry, Avogadro, and others subsequently stated the theoretical foundation for the physiology of O_2 and CO_2. Considering the recent scientific advances, respiration was defined as process, when energy was produced with the intake of oxygen and the release of carbon dioxide.

Later work showed that mitochondrial respiration is responsible for the O_2 consumption and CO_2 production observed by Spallanzani. This aspect of respiration is often called **internal respiration** or oxidative phosphorylation. Respiratory physiology focuses on **external respiration,** the dual processes of (1) transporting O_2 from the atmosphere to the mitochondria and (2) transporting CO_2 from the mitochondria to the atmosphere. CO_2 transport is intimately related to acid-base homeostasis.

Traditionally the process of respiration is divided into three phases: (1) cellular respiration, (2) transport of respiratory gases and (3) ventilation of the gas exchange organs (breathing) [6].

The main goals of respiration are oxygen uptake and elimination of carbon dioxide. Secondary goals include acid-base buffering, hormonal regulation, and host defense. To achieve the goals of respiration, three main functional components of the respiratory system are used: (1) mechanical structures, (2) membrane gas exchanger and (3) regulatory system (network of chemical and mechanical sensors throughout the circulatory and respiratory systems). All three components are tightly integrated, and dysfunction of one can lead to respiratory distress or failure [7].

Different kinds of neural receptors are present in the respiratory airways, lungs and pulmonary blood vessels:

- Changes in lung volume are perceived by pulmonary stretch receptors and muscle spindles.

- Rapidly adapting irritant receptors respond both to changes in lung volume and to the presence of chemicals such as histamine, prostaglandins, and exogenous noxious agents.

These receptors send the signals to the respiratory centers in the brain via the vagus nerve. On the other hand, the respiratory centers affect the breathing pattern by increasing the respiratory rate and/or stimulating cough, bronchoconstriction, and/or mucus production [8].

Input from these neural receptors likely accounts for the hyperventilation and hypocapnia that can occur in patients with pulmonary fibrosis even when hypoxemia is reversed by the administration of oxygen. Hyperventilation may occur by this mechanism in patients with such problems as asthma, interstitial lung disease, pulmonary edema, pneumonia, and pulmonary embolism.

Respiratory centers in the medulla receive stimulatory input from central respiratory pacer cells, central and peripheral chemoreceptors, upper airway receptors, other areas of the brain, and volitional pathways and integrate these signals into a combined output to respiratory muscles to modulate breathing frequency, inspiratory time, and expiratory time.

2.1 Developmental considerations

Significant changes occur in respiratory physiology during the transition from infancy to childhood, with the development of chest wall structures and maturation of the airways and lung parenchyma. Infancy is a time of rapid changes of central nervous system, neural respiratory control, as well as developmental plasticity and vulnerability. Rib cage geometry becomes more adult-like by about 3 years of age. The high chest wall compliance of the newborn decreases with age and becomes approximately equal to lung compliance, as in adults, by the second year of life, resulting in higher resting lung volume. These changes are important to recognize in the clinical setting because infants are more vulnerable to many disease states due to higher chest wall compliance, immature control of respiration, and increased airway resistance [9].

2.2 Nonrespiratory functions of the respiratory system

The main function of the respiratory system is gas exchange. However, the lung performs several nonrespiratory tasks. These functions include: its own defense against inspired particulate matter, the storage and filtration of blood for the systemic circulation, the handling of vasoactive substances in the blood, and the formation and release of substances used in the alveoli or circulation.

There are several ways in which respiratory system prevents from invading pathogens compromising the upper airway. Mucociliary clearance provides a strong physical barrier, several proteins, antimicrobial peptides, and reactive oxygen species, such as nitric oxide, play a significant role in preventing infection. The upper respiratory tract also has the ability to sense invading pathogens through Toll-like receptors and taste receptors that initiate immune responses [10, 11].

3. Conclusion

Considering all aspects of respiratory physiology, we can assume that with every breath we take, we provide the organism with a power to maintain homeostasis. The body carefully controls endless list of vital parameters and is always ready to adapt to changing circumstances.

Application of essential principles of physiology and staying up to date to constantly changing knowledge in the field is a bridge to treating a patient. Remembering physiological and pathophysiological mechanisms and their impact on health and disease will help the practitioners throughout their professional careers.

Author details

Ketevan Nemsadze[1,2,3,4]

1 Corresponding Member, Georgian National Academy of Sciences, Georgia

2 Fellow, Academy of Breastfeeding Medicine, FABM, USA

3 President, Georgian Academy of Pediatrics, Georgia

4 Full Professor, David Tvildiani Medical University, Georgia

*Address all correspondence to: ketinemsadze@gmail.com

IntechOpen

References

[1] Guyton and Hall Textbook of Medical Physiology 13th edition 2016 p. vii

[2] Guyton and Hall Textbook of Medical Physiology 13th edition 2016 p. 3

[3] Medical Physiology by Walter Boron Emile L Boulpaep 3rd edition 2017

[4] Ganong's Review o Medical Physiology 26th edition 2019 p. 17

[5] Central Role of the Brain in Stress and Adaptation B.S. McEwen, in Stress: Concepts, Cognition, Emotion, and Behavior, 2016 https://www.sciencedirect.com/topics/biochemistry-genetics-and-molecular-biology/homeostasis

[6] http://paulinesbiology.weebly.com/home/the-importance-of-respiration

[7] Respiratory Distress Juliann Lipps Kim, in Comprehensive Pediatric Hospital Medicine, 2007 https://www.sciencedirect.com/topics/neuroscience/respiratory-physiology

[8] Control of ventilation https://www.uptodate.com/contents/control-of-ventilation?csi=8d8a11eb-44a4-4d6d-ab75-487f6bbeac78&source=contentShare

[9] https://www.sciencedirect.com/topics/neuroscience/respiratory-physiology

[10] Pulmonary Physiology by Michael G. Levitzky, PhD Seventh Edition p.217

[11] Early local immune defenses in the respiratory tract Akiko Iwasaki1,2, Ellen F. Foxman1,3, and Ryan D. Molony1 Nat Rev Immunol. 2017 January ; 17(1): 7-20. doi:10.1038/nri.2016.117.

Respiratory Physiology and Exercise

Chapter 2

The Respiratory System during Intermittent-Sprint Work: Respiratory Muscle Work and the Critical Distribution of Oxygen

Ramón F. Rodriguez, Robert J. Aughey and François Billaut

Abstract

In healthy individuals at rest and while performing moderate-intensity exercise, systemic blood flow is distributed to tissues relative to their metabolic oxygen demands. During sustained high-intensity exercise, competition for oxygen delivery arises between locomotor and respiratory muscles, and the heightened metabolic work of breathing, therefore, contributes to limited skeletal muscle oxygenation and contractility. Intriguingly, this does not appear to be the case for intermittent-sprint work. This chapter presents new evidence, based on inspiratory muscle mechanical loading and hypoxic gas breathing, to support that the respiratory system of healthy men is capable of accommodating the oxygen needs of both locomotor and respiratory muscles when work is interspersed with short recovery periods. Only when moderate hypoxemia is induced, substantial oxygen competition arises in favour of the respiratory muscles. These findings extend our understanding of the relationship between mechanical and metabolic limits of varied exercise modes.

Keywords: blood flow, hyperpnoea, metaboreflex, oxygen uptake, hyperventilation, muscle fatigue

1. Introduction

Blood flow to contracting skeletal muscles closely matches their metabolic rate [1, 2]. In humans, it has been robustly demonstrated that there is a positive linear relationship between the rate of oxygen uptake (VO_2) in the quadriceps muscles and blood flow through the femoral artery [1], which ensures there is a match between oxygen (O_2) supply and demand for the exercising muscles. Blood flow is directed to areas in need by adjusting vasoconstriction in the relatively inactive regions and vasodilatation in the active locomotor muscles [2–4]. During high-intensity and maximal exercise, the accompanying increase in cardiac output is almost exclusively devoted to the working skeletal muscle [5], whereas blood flow to the splanchnic, renal and inactive skeletal muscle tissue beds can fall by ≈70% from resting values [6, 7]. It is likely that multiple biological factors contribute to biological redundancy in the system [8]. However, there does appear to be a limit to systemic vasodilation, a procreative mechanism to maintain arterial blood pressure and ensure adequate

O_2 supply to vital organs [4, 9–11]. Additionally, when the metabolic demands of multiple muscle groups are high, and cardiac output is nearing maximal flow rates, competition for available blood flow can arise between muscle groups. One such example is the interplay between limb locomotor musculature and the respiratory muscles.

The respiratory muscles perform work to overcome the elastic recoil of the lungs and chest, resistance from turbulent and viscous airflow through the respiratory tract and tissue deformation [12]. As pulmonary ventilation (V_E) rises, there is an exponential increase in the work being performed by the respiratory muscles [12, 13]. This ventilation-induced rise in work of breathing is caused by two factors; (1) dynamic hyperinflation to accommodate greater expiratory flow rates [14], and (2) progressive increase in the contribution of the expiratory muscles to breathing [15]. As the lungs and chest are progressively stretched to accommodate the increasing volume of inhaled air and end-expiratory lung volume is reduced, the contribution of elasticity in these tissues to the work of breathing increases [16, 17]. Accompanying the changes in work of breathing with V_E, there is a certain O_2 cost of exercise hyperpnoea [13, 18]. By mimicking the ventilation pattern (respiratory frequency and tidal volume) obtained during exercise while at rest, it is possible to estimate the proportion of whole-body VO_2 that is devoted to the respiratory muscles. During moderate exercise, the O_2 cost of breathing accounts for 3–6% of the total whole-body VO_2. During high-intensity exercise, the relative contribution of exercise hyperpnoea to whole-body VO_2 is estimated increases to 10–15% and can become a limiting factor of exercise capacity [2, 9, 19–21].

2. Consequences of sustained respiratory muscle work during continuous exercise

The work of breathing associated with high-intensity and maximal exercise is responsible for stealing a considerable portion of whole-body VO_2, which creates an environment where the locomotor and respiratory muscles compete for O_2 delivery [2, 3]. As such, respiratory muscle work, fatigue and metaboreflex are interrelated and suggested to contribute to the development of locomotor muscle fatigue, limiting one's capacity to sustain high-intensity exercise [9, 22].

An inverse relationship exists between the work of breathing and leg O_2 uptake during maximal exercise [3]. To reduce the work of breathing, proportional assist ventilation (PAV) can be used to generate inspiratory pressure proportional to the effort of the patient/subject. Conversely, to elevate inspiratory muscle work, a mesh screen can be placed over the inspiratory line, or the aperture of an inspiratory port can be reduced. In one such study employing these techniques, subjects exercised at a workload sustainable for 2.5–3 min at, or near a work rate corresponding to the attainment of VO_{2max}. The work of breathing was attenuated by 60% with PAV and increased by 95% with inspiratory loading, compared to control during the exercise bout. Elevating the work of breathing had a negligible effect on whole-body VO_2. Moreover, both leg blood flow and VO_2 fell compared to control exercise which coincided with an increase in leg vascular resistance. These data suggest that cardiac output did not increase to accommodate the additional muscular work [2]. It is likely that blood flow was redistributed to the respiratory muscles to support the heightened metabolic activity at the expense of the locomotor muscles [3]. When the respiratory muscles were unloaded with PAV, there was a slight increase in limb blood flow which corresponded with an increase in leg VO_2. Thus, by reducing the metabolic demands of the respiratory muscles, O_2 delivery to the lower limbs can be improved. Taken together, these data indicate that the 'normal' work of breathing incurred during high-intensity exercise may actually be a limiting factor of O_2 transport to the locomotor muscle during high-intensity exercise [3].

Exercise intensity also plays an important role in the competition between the locomotor and respiratory muscles for available O_2. While exercising at submaximal work rates (50–75% VO_{2max}), there is a small but significant increase in whole-body VO_2 in response to an elevated work of breathing [23]. Since VO_2 responded proportionally to the changes in inspiratory muscle work, is has been concluded that there is enough capacity in cardiac output to increase and meet the demands of additional muscular work during submaximal exercise [3]. It is only during high-intensity exercise when cardiac output approaches maximal flow rates that competition for available blood flow begins to develop [3].

Since the high work of breathing during high-intensity exercise (>80% VO_{2max}) seems to have a limiting effect on locomotor muscle blood flow, the rate of development of peripheral fatigue is likely affected too. To examine exercise-induced quadriceps muscle fatigue, supra-maximal femoral nerve stimulation can be used to provide an objective measure of muscle force-generating capacity [24]. In one such example, peripheral muscle fatigue was assessed after exercise at a work rate corresponding to the attainment of 92% of VO_{2max} [25]. On one occasion, subjects exercised to volitional exhaustion (13.2 min). On a separate visit, exercise at the same work rate and duration was repeated while the respiratory muscles were unloaded using PAV (56% reduction of inspiratory muscle work). Following the completion of exercise, quadriceps muscle fatigue was 8% greater when subjects were not using breathing assistance. To examine how a heightened work of breathing affects peripheral fatigue, exercise was repeated with inspiratory loading (80% increase of inspiratory muscle work) to exhaustion (7.9 min). Following the termination of exercise, the force-generating capacity of the quadriceps was 8% lower when performed with inspiratory loading compared to control [25]. These data robustly demonstrate that peripheral fatigue can be manipulated by altering the work for breathing, which suggests that respiratory muscle work is a limiting factor of high-intensity exercise [2, 3].

It is usually unlikely for healthy humans to experience inspiratory loading during exercise. However, exposure to (simulated) altitude is a more common environmental condition that will increase the work of breathing compared to normoxia via stimulation of pulmonary ventilation [26, 27]. To examine the relationship between hypoxia-induced elevated work of breathing and peripheral fatigue, subjects exercised at a constant work rate ($\approx$273 W) corresponding to 82% of VO_{2max} in simulated altitude (fraction of inspired oxygen, F_iO_2 = 0.15) to exhaustion [28]. Exercise was then repeated at the same work rate for an identical duration in normoxia ($\approx$273 W for 8.6 min). Compared to hypoxia, inspiratory muscle work was 36% less when exercising in normoxia and induced a lesser reduction in quadriceps force generation (normoxia −16% vs., hypoxia −30%). To isolate the effects of the work of breathing on peripheral muscle fatigue, subjects repeated both exercise trials (normoxia and hypoxia) using PAV. Inspiratory muscle work was nearly identical during exercise between normoxia and hypoxia with PAV, and the reduction in hypoxia-induced peripheral fatigue was attenuated relative to normoxia (normoxia −15% vs., hypoxia −22%). Combined, these data demonstrate that the development of quadriceps fatigue is accelerated in hypoxia in part due to heightened inspiratory muscle work. Moreover, this occurs at a work rate and exercise duration at which inspiratory muscle work usually does not affect quadriceps fatigue [28]. Sustained exercise ≥90% VO_{2max} and the accompanying work of breathing may have to reach a given threshold to elicit meaningful changes in quadriceps fatigue [25].

3. Factors influencing respiratory muscle work during repeated-sprint exercise

Repeated-sprints are characterised by brief 'all-out' exercise bouts of 4–15 s, separated by incomplete recovery periods of 14–30 s [29, 30]. Performance in a repeat-sprint context is therefore represented as the ability to reproduce power output after a previous bout of maximal exercise [31]. Over the course of a repeated-sprint series, there is a progressive decline in total mechanical work performed in each successive sprint. The rate of performance decline is also typically accelerated in low O_2 environments [32]. Initial sprint performance is largely determined by muscular strength and power production [33], whereas the ability to resist fatigue and maintain performance is underpinned by aerobic capacity and the ability to deliver O_2 to the locomotor muscles in the recovery periods between sprint recovery periods [34, 35]. Below, we outline the work load-induced physiological factors known to the work of breathing during intense intermittent exercise.

3.1 Metabolic determinates of repeated-sprint exercise

Resting intramuscular stores of ATP are limited to $\approx$20–25 mmol·kg^{-1} of dry muscle weight, which during a sprint, can only provide energy for 1–2 s [36, 37]. As resting ATP stores become depleted, three major energy systems are responsible for ATP resynthesis. Rapid resynthesis is achieved through phosphocreatine (PCr) degradation [36]. Anaerobic glycolysis also has a large involvement in sprint metabolism [36]. Though as sprints are repeated, the relative contribution of anaerobic glycolysis towards ATP resynthesis declines [36, 37]. Conversely, aerobic metabolism has a very small role in isolated sprit performance ($\approx$10% of total ATP production), which increases as sprints are repeated [37, 38].

Intramuscular PCr is especially important for the rapid resynthesis of ATP during explosive activities via the reversible PCr-creatine kinase pathway [39–41]. In the presence of the enzyme creatine kinase, adenosine diphosphate (ADP) is converted to ATP through the dephosphorylation of PCr to form creatine (Cr). It is estimated that during a single 6-s sprint, 50% of anaerobic ATP production is derived predominantly through PCr degradation [36]. The remaining anaerobic energy contribution during an isolated sprint is supported mainly by glycolysis (44%), and in minority by intramuscular ATP stores (6%). When sprints are repeated, the relative contribution of PCr to anaerobic ATP resynthesis increases. By the tenth 6-s sprint (each separated by 30 s passive rest), PCr degradation is estimated to account for 80% of the total anaerobic energy contribution [36]. However, intramuscular PCr stores are limited to $\approx$80 mmol·kg^{-1} of dry muscle weight, and after only a single 6-s sprint, stores are reduced $\approx$50% from baseline [36, 42]. When multiple sprints are performed, PCr depletion can be up to 75% after 5 repetitions [42], and 84% after 10 [36]. Since PCr degradation has such a large contribution to ATP resynthesis, the recovery of intramuscular stores PCr is critically important to the restoration of power output [43].

The capacity to recover PCr is limited in a multiple sprint series, largely constrained by the short recovery periods between sprints. The rate of PCr resynthesis follows an initial fast phase, followed by a second longer slow component [44]. After a single 6-s sprint, approximately 70% of PCr replenishment is achieved in the first 30 s of passive rest [42]. But as sprints are repeated and muscle stores are further depleted, PCr can only recover to 50% of resting stores after just five repetitions. When rest is extended post a repeat-sprint series, only 80% of PCr is recovered after 3 min [42], and 85% after 6 min of passive rest [45]. Though PCr degradation is an anaerobic process, PCr resynthesis is an aerobic process and is sensitive to O_2 availability [43, 44, 46, 47]. When breathing a hypoxic gas mixture

(F_1O_2 = 0.10), the rate of PCr resynthesis has been demonstrated to be attenuated by 23% [46]. While breathing a hyperoxic gas (F_1O_2 = 1.00) enhances recovery by 20% compared with normoxia, which suggests that under normal exercise conditions PCr resynthesis is limited by O_2 availability [46]. Therefore, if the work of breathing is high enough to limit locomotor muscle O_2 delivery, PCr resynthesis in repeated-sprint exercise may be impaired.

The energy debt created by the rapid decrease in muscle PCr during a single sprint is met by a sizable contribution of anaerobic glycolysis to ATP resynthesis. Approximately 44% of ATP resynthesis is derived from anaerobic glycolysis during a single 6-s sprint [36]. However, the relative contribution of anaerobic metabolism declines as sprints are repeated [48]. By the tenth sprint, Gaitanos, Williams [36] estimated that glycolysis was only responsible for 16% of total anaerobic ATP production. Moreover, in four of the seven subjects, it was estimated to be zero (range 0–23.1 mmol ATP·kg^{-1} of dry muscle weight). Many mechanisms play a role in the relative decrease in anaerobic glycolysis during multiple-sprint work. The most likely being the progressive depletion of muscle glycogen that is associated with high-intensity activity [49].

The aerobic contribution to an isolated sprint is minimal since the maximal rate of ATP resynthesis is far below the requirements of maximal sprint work [39]. In an isolated sprint, aerobic metabolism is responsible for ≈10% of total energy production [37, 48]. But as sprints are repeated, the relative increase in aerobic metabolism to total ATP turnover rate rises to compensate for reduced energy supply from anaerobic pathways [38]. Following five 6-s sprints, it is estimated the aerobic energy contribution rises to ≈40% of total ATP production [48]. The remaining 60% is derived from anaerobic pathways, predominantly PCr degradation [36, 42]. Pulmonary VO_2 can fluctuate between 70 and 100% of VO_{2max} from sprint to recovery periods in the latter stages of a repeat-sprint series [50]. When no external work is being performed (i.e., passive rest) during the recovery period between sprints, the elevated VO_2 above baseline is representative of lactate metabolism, removal of inorganic phosphate, and most importantly PCr resynthesis [40, 51].

Aerobic metabolism may have a limited role in ATP formation during multiple sprint work [38, 48], but is fundamental to PCr resynthesis between sprints. Compartment specific creatine kinase isozymes are located in the cytosol and mitochondrial intermembrane space, and are associated with either the ATP-consuming or -delivering process, respectively [40, 41]. In the PCr shuttle system, mitochondrial creatine kinase mediates the reaction between creatine and ATP formed by oxidative metabolism, to generate PCr and ADP [40]. Therefore, the rate at which the mitochondria can generate ATP through oxidative phosphorylation, will dictate PCr resynthesis. A positive correlation between aerobic fitness and maintaining repeat-sprint performance exists [31, 34, 52, 53]. It is likely that improvements in mitochondria function and content, that are associated with exercise training [54], underpin the correlation between aerobic fitness and repeated-sprint ability. Additionally, muscle O_2 availability between sprint efforts likely affects mitochondrial oxidative phosphorylation, which would explain the connection between PCr resynthesis and O_2 availability [43, 46]. Therefore, the ability to deliver O_2 to the locomotor muscles during rest periods between sprints is critical to maintaining maximal sprint performance [35, 47].

3.2 Muscle oxygenation and repeated-sprint exercise

Muscle O_2 availability during repeated-sprint exercise is critical for supporting PCr resynthesis, which underpins the capacity to maintain power out over a sprint series [36, 46]. Changes in local O_2 balance (delivery vs. consumption) can be measured in real-time with near-infrared spectroscopy (NIRS) [55]. The NIRS technology relies on the relative transparency of biological tissue to near-infrared light (650–950 nm), and light absorption of deoxyhaemoglobin and oxyhaemoglobin [56]. The concentration of deoxyhaemoglobin ([HHb]) and oxyhaemoglobin ([O_2Hb]) rises and falls, respectively, proportional to an increase in metabolic activity in the underlying tissue and display similar kinetics to pulmonary VO_2 [50, 57]. The analysis is typically focused on [HHb] since it is less sensitive to fluctuations in total haemoglobin, is assumed to reflect venous [HHb] and thus muscular oxygen extraction, and because [O_2Hb] is influenced by rapid blood volume and perfusion variations due to the skeletal muscle pump.

Because PCr resynthesis is achieved through oxidative processes [46, 58], the availability of muscle O_2 during rest periods is critically important for metabolic recovery. In maximal voluntary isometric handgrip exercise, reoxygenation rate measured as the rate change of [O_2Hb] during recovery was strongly correlated with the recovery of muscle PCr ($r^2 = 0.939$) [47]. Therefore, factors affecting muscle reoxygenation between sprint efforts will likely affect PCr resynthesis and repeated-sprint performance.

Vastus lateralis reoxygenation capacity can be attenuated by performing low-intensity activity (jogging/cycling) between sprint efforts [50, 59]. By reducing O_2 availability, the restoration of peak cycling power and peak running speed following periods of 'active' recovery is 3–7% lower compared to passive rest. The time to exhaustion is also lowered by performing 'active' recovery when performing 15-s sprints, repeated every 15 s (745 ± 171 s vs. 445 ± 79 s; −60%) [60]. Performing active recovery between sprints, muscle tissue reoxygenation is impaired through the constant O_2 uptake supporting the metabolic requirements of the active recovery. Therefore, PCr resynthesis is likely blunted because ATP from oxidative phosphorylation is devoted directly to maintain muscle contractions, rather than towards PCr resynthesis [41, 59].

The influence of limited reoxygenation on repeated-sprint ability has also been highlighted by manipulating the F_1O_2. When performing ten 10-s sprints with 30 s of passive rest and inspiring a hypoxic gas mixture ($F_1O_2 = 0.13$), reoxygenation was attenuated by 11% [35]. There was a $\approx 8\%$ reduction in total mechanical work in hypoxia compared to normoxia, and the reduction in work was strongly correlated with the attenuated muscle reoxygenation ($r = 0.78$; 90% confidence interval: 0.49, 0.91). Since PCr resynthesis has similar recovery kinetics to reoxygenation [47], it is likely that muscle PCr recovery was hindered by limited O_2 availability. Therefore, enhancing the capacity to reoxygenation the muscle between sprints is likely to have positive benefits for repeated-sprint ability.

There exists a positive relationship between aerobic fitness and repeated-sprint ability, which may in part be explained by superior reoxygenation capacity [31, 34, 52, 53]. After 8 weeks of endurance training, although the initial sprint performance is typically unaffected [61] (presumably because improvements in aerobic function do not support the anaerobic nature of an isolated sprint), muscle oxygenation was reported to be 152% higher prior to the commencement of the second sprint following training. Consequently, the decrement in performance within the subsequent sprint was attenuated by 26% [61]. It is likely that by improving O_2 delivery to the locomotor muscle, O_2 availability for oxidative phosphorylation was enhanced, and in turn, the phosphocreatine shuttle system [39, 40].

3.3. Heightened inspiratory muscle work

As described in Section 2, respiratory muscle work has been implicated as a limiting factor of limb O_2 perfusion during continuous exercise [9]. However, competition between locomotor and respiratory muscle for available cardiac output does not appear to be a significant limiting factor of performance during repeated-sprint exercise. In our recent work, we examined the influence of inspiratory muscle loading on oxygenation trends in repeated-sprint exercise [62]. Participants were asked to perform ten 10-s cycle ergometer sprints, each separated by 30 s of passive rest. Inspiratory loading was achieved by placing a plastic disk with a 10-mm opening over the inspiratory side of a two-way non-rebreathing valve (**Figure 1**). Inspiratory muscle force development (calculated as the integral of inspiratory mouth pressure, multiplied by respiratory frequency) was similar to others who have shown vastus grater lateralis muscle deoxygenation with inspiratory loading during exercise [63]. In response, whole-body VO_2 measured at the mouth was elevated by 4–5% during both the sprint and recovery phases (**Figure 2**) [62]. This occurred even though total sprint work was similar between the conditions. The elevation in VO_2 was likely driven by a heightened oxygen O_2 uptake by the respiratory muscle to accommodate for the additional inspiratory muscle work [23, 63]. Importantly, Tissue Saturation Index (TSI = $[O_2Hb] \div ([O_2Hb] + [HHb])$, expressed in %) measured at the sixth intercostal space, was comparable between the conditions, which suggests that O_2 supply to the respiratory muscles remained proportional to the metabolic activity. The change in inspiratory muscle work did not translate into compromised vastus lateralis oxygenation.

The intermittent nature of repeated sprints is likely a key mediating factor for which O_2 delivery can be maintained to both locomotor and respiratory

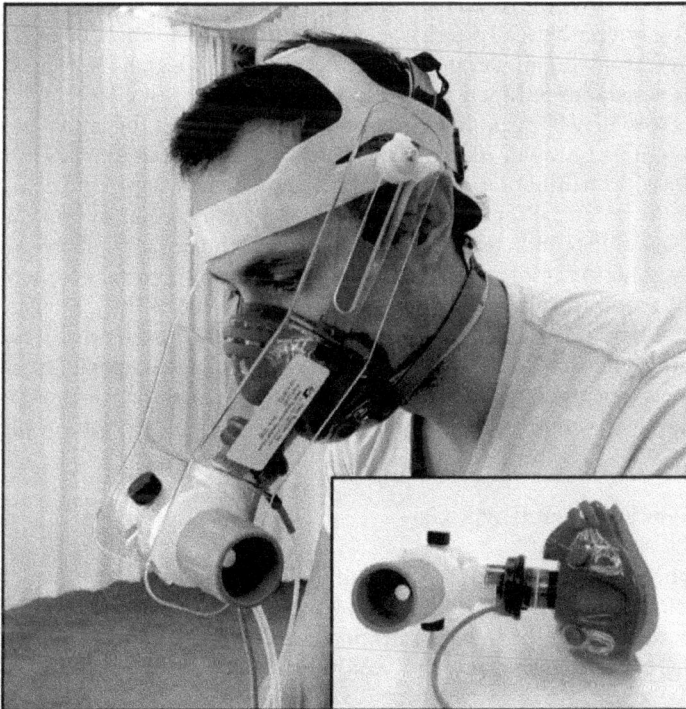

Figure 1.
Representation of how inspiratory loading was achieved [62]. A plastic disk with a 10-mm opening was placed over the inspiratory side of a two-way non-rebreathing valve (Hans Rudolph Inc., Kansas, United States of America) attached to the distal end of a bidirectional turbine and held in place by the internal ridge of a rubber tubing adaptor. A pressure transducer was attached to the saliva port of the non-rebreathing valve via Tygon tubing to assess inspiratory mouth pressure.

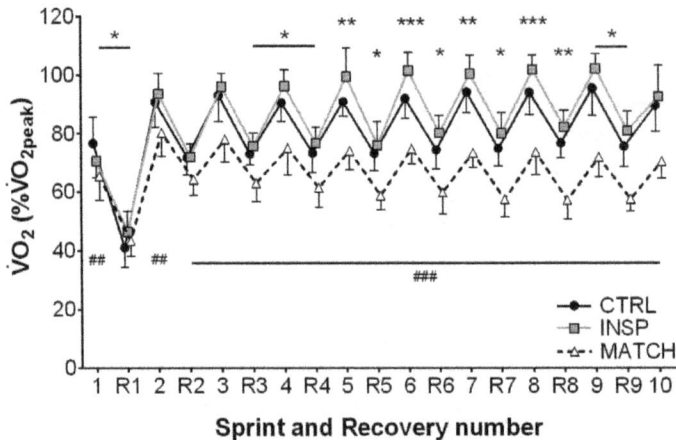

Figure 2.
Sprint and recovery pulmonary oxygen uptake (VO$_2$) expressed as a percentage of VO$_{2peak}$ for control (CTRL), inspiratory muscle loading (INSP), and worked matched (MATCH) exercise. The symbols represent comparisons between INSP and CTRL (*), INSP and MATCH (#). The number of symbols; one, two, and three denote likely, very likely and almost certainly respectively, that the chance of the true effect exceeds a small (−0.2 to 0.2) effect size. Values are presented as mean ± SD. Reproduced from Rodriguez et al. [62] under a Creative Commons Attribution 4.0 licence.

muscles. Others have demonstrated that the addition of an inspiratory load while exercising >95% VO$_{2max}$, results in a decrease in limb perfusion and O$_2$ delivery, mediated by sympathetically-activated vasoconstriction in the locomotor muscles [2, 3]. Whereas during moderate intensities (50–75% VO$_{2max}$), there is no change in vascular resistance or blood flow [23]. Even though repeated-sprint exercise can elicit >90% of VO$_2$ peak, it is not sustained throughout the entire protocol and can fluctuate between 70 and 90% of VO$_{2max}$ between sprint and recover phases [50, 62]. The fluctuation in metabolic demands between the phases likely minimises the potential for a competition for available cardiac output. Moreover, since VO$_2$ was able to increase, these data highlight the capacity of the cardiorespiratory system to rapidly adjust and meet the additional metabolic demands imposed by inspiratory loading even during severe exercise [64]. In instances where blood flow is impacted by additional respiratory muscle work, there is no compensatory increase in VO$_2$ [2, 3]. Therefore, having the capacity to increase VO$_2$ may be a crucial factor in maintaining O$_2$ supply to all active muscles during high-intensity exercise and, thereby, sustain prolonged periods of physical activity.

3.4 Acute environmental hypoxia

To further explore the role of O$_2$ availability in balancing the metabolic demands of the locomotor and respiratory muscles, we asked participants to exercise in an environment where the O$_2$ concentration had been reduced to 14.55% [65]. Participants completed the same protocol as previously described (ten 10-s sprints, 30 s of passive rest) while vastus lateralis and intercostal muscle oxygenation was assessed with NIRS. Surprisingly, there was no clear difference in repeated-sprint ability in hypoxia compared to normoxia. However, there was a clear reduction in vastus lateralis muscle oxygenation similar to previous research (**Figure 3**) [35, 66]. Ventilation patterns (respiratory frequency and inspiratory volume) and inspiratory pressure generation were similar between conditions. Therefore, the O$_2$ cost of

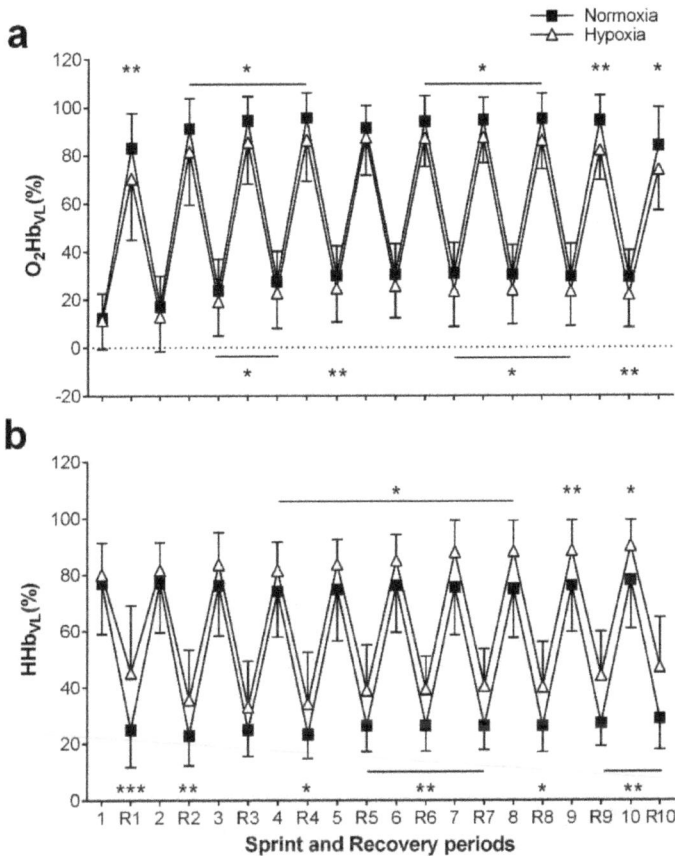

Figure 3.
Vastus lateralis muscle oxygenation trends during repeated-sprint exercise in normoxia and hypoxia expressed as a percentage of arterial occlusion. (a) Change of vastus lateralis oxyhaemoglobin (O_2Hb_{VL}); and (b) vastus lateralis deoxyhaemoglobin (HHb_{VL}). The number of symbols (*); one, two and three denote likely, very likely and most likely respectively, that the chance of the true effect exceeds a small (−0.2 to 0.2) effect size. Results are presented as mean ± SD. Reprinted from Rodriguez et al. [65], with permission from Elsevier.

exercise hyperpnoea was likely similar between conditions [67]. However, there was no clear difference in respiratory muscle oxygenation ([O_2Hb] and [HHb]) during exercise in hypoxia when compared to normoxia (**Figure 4**) [65]. Based on this new evidence, it appears that O_2 delivery is preferentially distributed to the respiratory muscles to maintain the metabolic function of the respiratory muscles.

Locomotor muscle O_2 availability during rest phases between sprints is a strong determining factor of metabolic recovery [36, 37, 44], and thus performance over multiple sprints [35]. Based on our early research, it seems that locomotor muscle O_2 availability is compromised in hypoxia in favour of the respiratory muscles. Others have reported exaggerated deoxygenation of the respiratory muscles in hypoxia during voluntary isocapnic hyperpnoea [68]. However, their hypoxia gas mixture (10% O_2) resulted in a lower arterial O_2 saturation of 82% (estimated via pulse oximetry) compared to the average 87% in subjects of the study discussed here [65]. A hypoxic threshold may exist where respiratory muscle O_2 delivery can be maintained close to the rate of that during exercise in normoxia. If arterial hypoxemia was greater, further desaturation of the respiratory muscles may have been detected. Amann et al. [28] have reported a link between inspiratory muscle

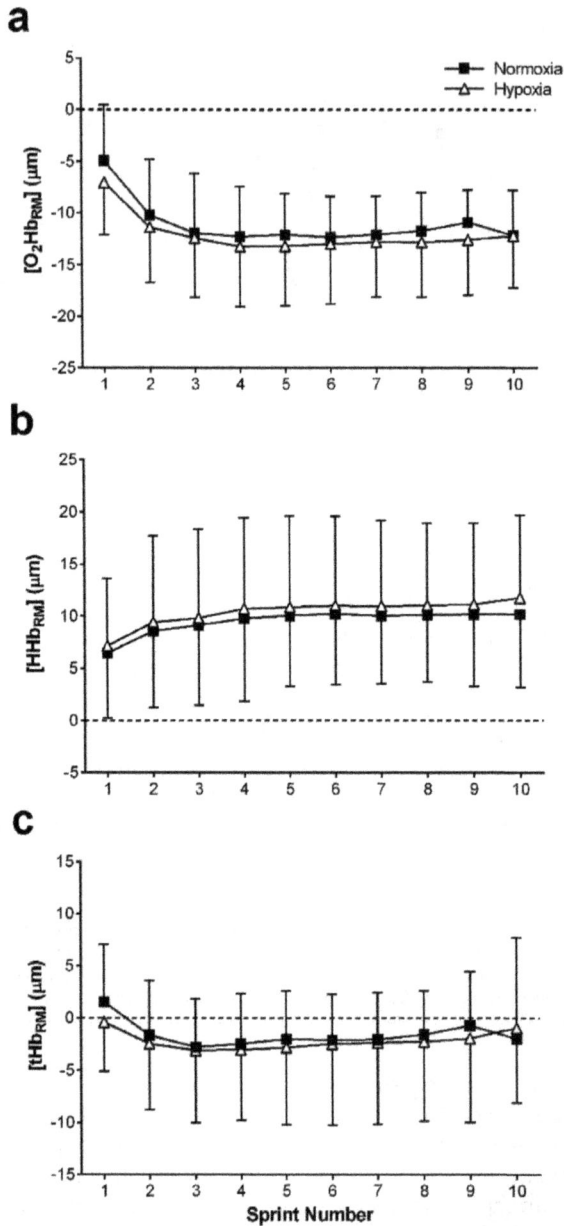

Figure 4.
Respiratory muscle oxygenation trends during repeated-sprint exercise in normoxia and hypoxia expressed as an absolute change from baseline (horizontal line). (a) Concentration change from baseline of respiratory muscle oxyhaemoglobin ([O_2Hb_{RM}]); (b) respiratory muscle deoxyhaemoglobin ([HHb_{RM}]); and (c) respiratory muscle total haemoglobin ([tHb_{RM}]). There was no clear effect of hypoxia on respiratory muscle oxygenation compared to normoxia. Results are represented as mean ± SD. Reprinted from Rodriguez et al. [65], with permission from Elsevier.

work in hypoxia and the development of quadriceps fatigue during high-intensity exercise. By reducing the work of breathing with PAV, the rate of fatigue developments can be attenuated [28]. Therefore, alleviating the O_2 cost of exercise hyperpnoea appears to be a pathway for enhancing limb O_2 delivery and exercise capacity in humans.

3.5 Respiratory muscle training

Aside from the structural characteristics of the pulmonary system, the relative strength of the respiratory muscles is likely to have a key role in the O_2 cost of exercise hyperpnoea. After 6-weeks of inspiratory muscle strength training, is has been demonstrated that ventilation O_2 efficiency can be enhanced [19]. Specific training targeting the inspiratory muscles (inspiratory muscle training, IMT) typically consists of inspiring against a closed valve set to open at ≈50% of an individual's maximal inspiratory mouth pressure, repeated 30 times twice per day. Strengthening the inspiratory muscles has translated to reduced O_2 cost of voluntary hyperpnoea, attenuated exercise-induced respiratory muscle fatigue, attenuated vastus lateralis and respiratory muscle deoxygenation, and improved exercise capacity [19, 67–70]. However, respiratory muscle fatigue has not been clearly demonstrated for multiple-sprint work, and therefore ergogenic benefits of respiratory muscle training for improving repeated-sprint alibility may be limited [71]. Nevertheless, evidence that IMT provides some benefit towards maintaining repeated-sprint performance exists, though the mechanisms are unclear [72, 73].

After a 6-week period of IMT, repeated-sprint ability was assessed in a group of recreational sprint sports players (soccer, rugby, field hockey and basketball) [72]. Performance was assessed during fifteen 20-m sprints, which participants were allowed a maximum of 30 s rest. Following the intervention, there were no clear changes in sprint times. However, self-selected recovery time was lessened by 6.9% (range: −0.9 to 14.5%). Strengthening the inspiratory muscles presumably reduced the O_2 cost of exercise hyperpnoea and blunted the respiratory muscle metaboreflex, which would, in turn, reduce O_2 competition between locomotor and respiratory muscles [9, 19, 74]. Through minimising O_2 competition, it is likely that the quality of metabolic recovery was enhanced with IMT, so that subjects could maintain performance with less rest between sprints [72]. But since there were no measurements of muscle oxygenation, it is difficult to separate potential changes in O_2 delivery from reduced feelings of dyspnoea that is associated with respiratory muscle training [69, 72].

The effectiveness of IMT on repeat-sprint ability and run time to exhaustion at 100% of the speed obtained during a maximal incremental exercise test has also been assessed in a group of professional female soccer players [73]. Repeated-sprint ability was assessed with six 40-m sprints (20 m + 180° turn +20 m) with 20 s of passive rest between sprints. Vastus lateralis and intercostal muscle oxygenation was only examined during the time-to-exhaustion trials. There was no significant difference between the groups in repeated-sprint ability ($P > 0.05$). However the effect size for performance decrement was slightly larger in the IMT group post intervention (Cohen's $d = 0.84$ vs. 0.16). Similar, both placebo and experimental groups improved time to exhaustion with no significant difference between groups, but the effect size in the IMT group was larger (Cohen's $d = 0.74$ vs. 0.46). Specific training of the respiratory muscles, therefore, may only provide negligible/small performance benefits beyond professional soccer periodised training. Performance benefits were partly attributed to a blunted increase in respiratory muscle [HHb], with a concurrent increase in vastus lateralis [O_2Hb] [73]. In terms of the athlete's ability to preserve repeat-sprint performance, the IMT group also showed the greatest improvement in the capacity to maintain sprint time over multiple sprints. The blunted respiratory muscle metaboreflex in the exhaustion test may have also occurred during the repeated-sprint test. However, without muscle oxygenation measurements during the sprint trials, it is unclear if there were any changes to O_2 availability after training.

The few studies demonstrating enhanced repeated-sprint performance following IMT [72, 73] support the notion that respiratory muscle work plays a negative effect on high-intensity intermittent exercise. Training the respiratory muscles can reduce the O_2 cost of exercise hyperpnoea [19], and attenuate blood flow competition between the locomotor and respiratory muscles [70, 74]. However, there remains a very limited understanding of the role exercise hyperpnoea plays during repeated-sprint exercise. Research still needs to answer if the enhanced repeated-sprint ability following respiratory muscle training is derived from improved skeletal muscle oxygenation kinetics.

3.6 Evidence for hyperventilation

Hyperventilation is demarcated when alveolar ventilation disproportionally rises relative to CO_2 production causing a decrease in the pressure of alveolar CO_2, and an increase in the pressure of alveolar O_2 [75]. Hyperventilation readily occurs during high-intensity exercise and can constrain a fall in arterial O_2 and pH [75, 76]. Though this was not directly examined in our research [62, 65], some evidence of hyperventilation occurring during repeated-sprint exercise was present. As depicted by the data of a representative subject (**Figure 5**), the partial pressure of end-tidal oxygen ($P_{ET}O_2$) and carbon dioxide ($P_{ET}CO_2$) rose and fell respectively from baseline over the course of the repeated-sprint protocol [62]. The wave-like pattern in $P_{ET}O_2$ and $P_{ET}CO_2$ appears to be linked to the phase of the protocol (sprint vs. rest) and occurs at exercise onset. This pattern is suggestive of a locomotor respiratory coupling, in which breathing frequency matches the cadence of locomotor exercise [77].

Further new evidence of hyperventilation comes from our recent hypoxia research [65]. Arterial hypoxemia is a potent stimulus of ventilation [26–28]. However, as we reported, there was no clear difference in either inspiratory volume, respiratory frequency or inspiratory mouth pressure during the repeated-sprint protocol. One may argue that participants were already operating at their

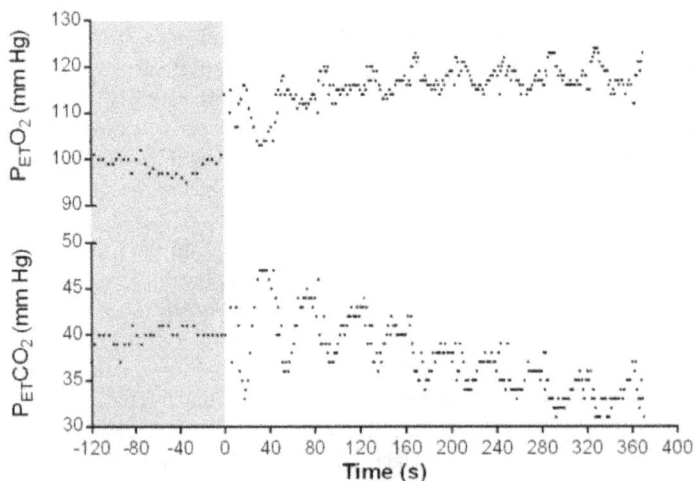

Figure 5.
Partial pressure of end-tidal oxygen ($P_{ET}O_2$) and carbon dioxide ($P_{ET}CO_2$) recorded on a breath-by-breath basis during repeated-sprint exercise. Data are from a single subject collected as part of the study by Rodriguez et al. [62] during the control exercise condition. The exercise protocol consisted of ten 10-s sprints, separated by 30 s of passive rest so that a sprint commenced every 40 s. The grey shaded area represents the 2-min baseline period observed prior to the commencement of warm-up.

upper limits of ventilation, and thus arterial hypoxemia could not have had an additive effect. Although appealing, such hypothesis requires additional work to determine the influencing factors of exercise hyperpnoea over a variety of sprint durations.

4. Conclusion

The findings of our research do not support heightened inspiratory muscle work as being a limiting factor in vastus lateralis muscle oxygenation in normoxia. The intermittent nature of repeated-sprint activity is likely a key mediating factor for which O_2 delivery can be maintained to both the locomotor and respiratory muscles. Moreover, reducing the relative intensity of exercise hyperpnoea through inspiratory muscle training shows limited benefits for enhancing repeated sprint ability. Inspiratory muscle work appears to play a more influential role under conditions of arterial hypoxemia. Our research showed that locomotor muscle oxygenation can be compromised through preferential O_2 delivery to the respiratory muscles. It is yet to be seen if inspiratory muscle training could be of benefit to exercise under these conditions.

Author details

Ramón F. Rodriguez[1], Robert J. Aughey[1] and François Billaut[1,2*]

1 Institute for Health and Sport, Victoria University, Melbourne, Australia

2 Department of Kinesiology, University Laval, Quebec, Canada

*Address all correspondence to: francois.billaut@kin.ulaval.ca

IntechOpen

References

[1] Andersen P, Saltin B. Maximal perfusion of skeletal muscle in man. The Journal of Physiology. 1985;**366**:233-249

[2] Harms CA et al. Effects of respiratory muscle work on cardiac output and its distribution during maximal exercise. Journal of Applied Physiology. 1998;**85**(2):609-618

[3] Harms CA et al. Respiratory muscle work compromises leg blood flow during maximal exercise. Journal of Applied Physiology. 1997;**82**(5):1573-1583

[4] Secher NH, Volianitis S. Are the arms and legs in competition for cardiac output? Medicine and Science in Sports and Exercise. 2006;**38**(10):1797-1803

[5] Joyner MJ, Casey DP. Regulation of increased blood flow (hyperemia) to muscles during exercise: A hierarchy of competing physiological needs. Physiological Reviews. 2015;**95**(2):549-601

[6] Rowell LB et al. Splanchnic vasomotor and metabolic adjustments to hypoxia and exercise in humans. American Journal of Physiology. 1984;**247**(2):H251-H258

[7] Poortmans JR. Exercise and renal function. Sports Medicine. 1984;**1**(2):125-153

[8] Joyner MJ, Wilkins BW. Exercise hyperaemia: Is anything obligatory but the hyperaemia? The Journal of Physiology. 2007;**583**(Pt 3):855-860

[9] Dempsey JA et al. Consequences of exercise-induced respiratory muscle work. Respiratory Physiology and Neurobiology. 2006;**151**(2-3):242-250

[10] Calbet JAL, Lundby C. Skeletal muscle vasodilatation during maximal exercise in health and disease. The Journal of Physiology. 2012;**590**(24):6285-6296

[11] Saltin B. Hemodynamic adaptations to exercise. The American Journal of Cardiology. 1985;**55**(10):D42-D47

[12] Otis AB, Fenn WO, Rahn H. Mechanics of breathing in man. Journal of Applied Physiology. 1950;**2**(11):592-607

[13] Aaron EA et al. Oxygen cost of exercise hyperpnea: Measurement. Journal of Applied Physiology. 1992;**72**(5):1810-1817

[14] Pellegrino R et al. Expiratory airflow limitation and hyperinflation during methacholine-induced bronchoconstriction. Journal of Applied Physiology. 1993;**75**(4):1720-1727

[15] Aliverti A et al. Human respiratory muscle actions and control during exercise. Journal of Applied Physiology. 1997;**83**(4):1256-1269

[16] Johnson BD et al. Exercise-induced diaphragmatic fatigue in healthy humans. The Journal of Physiology. 1993;**460**(1):385-405

[17] Guenette JA et al. Respiratory mechanics during exercise in endurance-trained men and women. The Journal of Physiology. 2007;**581**(3):1309-1322

[18] Dominelli PB et al. Oxygen cost of exercise hyperpnoea is greater in women compared with men. The Journal of Physiology. 2015;**593**(8):1965-1979

[19] Turner LA et al. Inspiratory muscle training lowers the oxygen cost of voluntary hyperpnea. Journal of Applied Physiology. 2012;**112**(1):127-134

[20] Aaron EA et al. Oxygen cost of exercise hyperpnea: Implications

for performance. Journal of Applied Physiology. 1992;**72**(5):1818-1825

[21] Harms CA et al. Effects of respiratory muscle work on exercise performance. Journal of Applied Physiology. 2000;**89**(1):131-138

[22] Romer LM, Polkey MI. Exercise-induced respiratory muscle fatigue: Implications for performance. Journal of Applied Physiology. 2008;**104**(3):879-888

[23] Wetter TJ et al. Influence of respiratory muscle work on VO_2 and leg blood flow during submaximal exercise. Journal of Applied Physiology. 1999;**87**(2):643-651

[24] Polkey MI et al. Quadriceps strength and fatigue assessed by magnetic stimulation of the femoral nerve in man. Muscle and Nerve. 1996;**19**(5):549-555

[25] Romer LM et al. Effect of inspiratory muscle work on peripheral fatigue of locomotor muscles in healthy humans. The Journal of Physiology. 2006;**571**(Pt 2):425-439

[26] Cibella F et al. Respiratory mechanics during exhaustive submaximal exercise at high altitude in healthy humans. The Journal of Physiology. 1996;**494**(3):881-890

[27] Cibella F et al. Respiratory energetics during exercise at high altitude. Journal of Applied Physiology. 1999;**86**(6):1785-1792

[28] Amann M et al. Inspiratory muscle work in acute hypoxia influences locomotor muscle fatigue and exercise performance of healthy humans. American Journal of Physiology—Regulatory, Integrative and Comparative Physiology. 2007;**293**(5):R2036-R2045

[29] Billaut F, Bishop D. Muscle fatigue in males and females during multiple-sprint exercise. Sports Medicine. 2009;**39**(4):257-278

[30] Glaister M. Multiple sprint work: Physiological responses, mechanisms of fatigue and the influence of aerobic fitness. Sports Medicine. 2005;**35**(9):757-777

[31] Bishop DJ, Edge J, Goodman C. Muscle buffer capacity and aerobic fitness are associated with repeated-sprint ability in women. European Journal of Applied Physiology. 2004;**92**(4):540-547

[32] Bowtell JL et al. Acute physiological and performance responses to repeated sprints in varying degrees of hypoxia. Journal of Science and Medicine in Sport. 2014;**17**(4):399-403

[33] Newman MA, Tarpenning KM, Marino FE. Relationships between isokinetic knee strength, single-sprint performance, and repeated-sprint ability in football players. Journal of Strength and Conditioning Research. 2004;**18**(4):867-872

[34] Gharbi Z et al. Aerobic and anaerobic determinants of repeated sprint ability in team sports athletes. Biology of Sport. 2015;**32**(3):207-212

[35] Billaut F, Buchheit M. Repeated-sprint performance and vastus lateralis oxygenation: Effect of limited O_2 availability. Scandinavian Journal of Medicine and Science in Sports. 2013;**23**(3):185-193

[36] Gaitanos GC et al. Human muscle metabolism during intermittent maximal exercise. Journal of Applied Physiology. 1993;**75**(2):712-719

[37] Parolin ML et al. Regulation of skeletal muscle glycogen phosphorylase and PDH during maximal intermittent exercise. American Journal of Physiology—Endocrinology and Metabolism. 1999;**277**(5):E890-E900

[38] Bogdanis GC et al. Contribution of phosphocreatine and aerobic metabolism to energy supply during repeated sprint exercise. Journal of Applied Physiology. 1996;**80**(3):876-884

[39] Baker JS, McCormick MC, Robergs RA. Interaction among skeletal muscle metabolic energy systems during intense exercise. Journal of Nutrition and Metabolism. 2010;**2010**:905612

[40] Guimaraes-Ferreira L. Role of the phosphocreatine system on energetic homeostasis in skeletal and cardiac muscles. Einstein (Sao Paulo). 2014;**12**(1):126-131

[41] Schlattner U, Tokarska-Schlattner M, Wallimann T. Mitochondrial creatine kinase in human health and disease. Biochimica et Biophysica Acta. 2006;**1762**(2):164-180

[42] Dawson B et al. Muscle phosphocreatine repletion following single and repeated short sprint efforts. Scandinavian Journal of Medicine and Science in Sports. 1997;7(4):206-213

[43] Sahlin K, Harris RC, Hultman E. Resynthesis of creatine phosphate in human muscle after exercise in relation to intramuscular pH and availability of oxygen. Scandinavian Journal of Clinical and Laboratory Investigation. 1979;**39**(6):551-558

[44] Harris RC et al. The time course of phosphorylcreatine resynthesis during recovery of the quadriceps muscle in man. Pflügers Archiv. 1976;**367**(2):137-142

[45] Mendez-Villanueva A et al. The recovery of repeated-sprint exercise is associated with PCr resynthesis, while muscle pH and EMG amplitude remain depressed. PLoS One. 2012;7(12):e51977

[46] Haseler LJ, Hogan MC, Richardson RS. Skeletal muscle phosphocreatine recovery in exercise-trained humans is dependent on O_2 availability. Journal of Applied Physiology. 1999;**86**(6):2013-2018

[47] Kime R et al. Delayed reoxygenation after maximal isometric handgrip exercise in high oxidative capacity muscle. European Journal of Applied Physiology. 2003;**89**(1):34-41

[48] McGawley K, Bishop DJ. Oxygen uptake during repeated-sprint exercise. Journal of Science and Medicine in Sport. 2015;**18**(2):214-218

[49] Balsom PD et al. High-intensity exercise and muscle glycogen availability in humans. Acta Physiologica Scandinavica. 1999;**165**(4):337-345

[50] Buchheit M et al. Muscle deoxygenation during repeated sprint running: Effect of active vs. passive recovery. International Journal of Sports Medicine. 2009;**30**(6):418-425

[51] Gaesser GA, Brooks GA. Metabolic bases of excess post-exercise oxygen consumption: A review. Medicine and Science in Sports and Exercise. 1984;**16**(1):29-43

[52] da Silva JF, Guglielmo LG, Bishop DJ. Relationship between different measures of aerobic fitness and repeated-sprint ability in elite soccer players. Journal of Strength and Conditioning Research. 2010;**24**(8):2115-2121

[53] Tomlin DL, Wenger HA. The relationship between aerobic fitness and recovery from high intensity intermittent exercise. Sports Medicine. 2001;**31**(1):1-11

[54] Bishop DJ, Granata C, Eynon N. Can we optimise the exercise training prescription to maximise improvements in mitochondria function and content? Biochimica et Biophysica Acta—General Subjects. 2014;**1840**(4):1266-1275

[55] Ferrari M, Muthalib M, Quaresima V. The use of near-infrared spectroscopy in understanding skeletal muscle physiology: Recent developments. Philosophical transactions Series A, Mathematical, Physical, and Engineering Sciences. 2011;**369**(1955):4577-4590

[56] Alhemsi H, Zhiyun L, Deen MJ. Time-resolved near-infrared spectroscopic imaging systems. In: BenSaleh MS, Qasim SM, editors. Saudi International Electronics, Communications and Photonics Conference (SIECPC); 27-30 April 2013: The Institute of Electrical and Electronics Engineers; 2013. p. 1-6

[57] Grassi B et al. Blood lactate accumulation and muscle deoxygenation during incremental exercise. Journal of Applied Physiology. 1999;**87**(1):348-355

[58] Hogan MC, Richardson RS, Haseler LJ. Human muscle performance and PCr hydrolysis with varied inspired oxygen fractions: A 31P-MRS study. Journal of Applied Physiology. 1999;**86**(4):1367-1373

[59] Ohya T, Aramaki Y, Kitagawa K. Effect of duration of active or passive recovery on performance and muscle oxygenation during intermittent sprint cycling exercise. International Journal of Sports Medicine. 2013;**34**(7):616-622

[60] Dupont G et al. Passive versus active recovery during high-intensity intermittent exercises. Medicine and Science in Sports and Exercise. 2004;**36**(2):302-308

[61] Buchheit M, Ufland P. Effect of endurance training on performance and muscle reoxygenation rate during repeated-sprint running. European Journal of Applied Physiology. 2011;**111**(2):293-301

[62] Rodriguez RF et al. Muscle oxygenation maintained during repeated-sprints despite inspiratory muscle loading. PLoS One. 2019;**14**(9):e0222487

[63] Turner LA et al. Inspiratory loading and limb locomotor and respiratory muscle deoxygenation during cycling exercise. Respiratory Physiology and Neurobiology. 2013;**185**(3):506-514

[64] Gleser MA, Horstman DH, Mello RP. The effect on Vo_{2max} of adding arm work to maximal leg work. Medicine and Science in Sports. 1974;**6**(2):104-107

[65] Rodriguez RF et al. Respiratory muscle oxygenation is not impacted by hypoxia during repeated-sprint exercise. Respiratory Physiology and Neurobiology. 2019;**260**:114-121

[66] Smith KJ, Billaut F. Influence of cerebral and muscle oxygenation on repeated-sprint ability. European Journal of Applied Physiology. 2010;**109**(5):989-999

[67] Dominelli PB et al. Precise mimicking of exercise hyperpnea to investigate the oxygen cost of breathing. Respiratory Physiology and Neurobiology. 2014;**201**:15-23

[68] Katayama K et al. Hypoxia exaggerates inspiratory accessory muscle deoxygenation during hyperpnoea. Respiratory Physiology and Neurobiology. 2015;**211**:1-8

[69] Downey AE et al. Effects of inspiratory muscle training on exercise responses in normoxia and hypoxia. Respiratory Physiology and Neurobiology. 2007;**156**(2):137-146

[70] Turner LA et al. The effect of inspiratory muscle training on respiratory and limb locomotor muscle deoxygenation during exercise

with resistive inspiratory loading.
International Journal of Sports
Medicine. 2016;(36):598-606

[71] Minahan C et al. Repeated-sprint
cycling does not induce respiratory
muscle fatigue in active adults:
Measurements from the Powerbreathe®
inspiratory muscle trainer. Journal
of Sports Science and Medicine.
2015;**14**(1):233-238

[72] Romer LM, McConnell AK,
Jones DA. Effects of inspiratory muscle
training upon recovery time during
high intensity, repetitive sprint
activity. International Journal of Sports
Medicine. 2002;**23**(5):353-360

[73] Archiza B et al. Effects of
inspiratory muscle training in
professional women football players:
A randomized sham-controlled
trial. Journal of Sports Sciences.
2018;**36**(7):771-780

[74] Witt JD et al. Inspiratory muscle
training attenuates the human
respiratory muscle metaboreflex.
The Journal of Physiology.
2007;**584**(3):1019-1028

[75] Forster HV, Haouzi P, Dempsey JA.
Control of breathing during exercise.
Comprehensive Physiology.
2012;**2**(1):743-777

[76] Whipp BJ, Ward SA. Determinants
and control of breathing during
muscular exercise. British Journal of
Sports Medicine. 1998;**32**(3):199-211

[77] Bernasconi P, Kohl J. Analysis of
co-ordination between breathing and
exercise rhythms in man. Journal of
Physiology. 1993;**471**(1):693-706

Chapter 3

The Main Clinical Indicators of Sarcopenia in Patients with Chronic Respiratory Disease: Skeletal Muscle Dysfunction Approach

Esmaeil Alibakhshi, Raffaele Fiorillo, Luis Lores Obradors, Ana María Sánchez-Laforga, Amparo Villar Cánovas, Mostafa Ghanei and Ali Qazvini

Abstract

Patients with chronic respiratory diseases (CRDs) have a disorder in muscle structure and function, but their function increases with physical progress and decreases the risk of general, and muscular weakness are more likely to develop sarcopenia. We randomly selected patients (N = 38) with mean age of 72 ± 1.0 years old men and women elderly with chronic respiratory diseases such as asthma, COPD, bronchiectasis and obesity with dyspnea score ≥ 2 in MRC index. All patients after receiving research information and signing informed consent have gone through performing clinical assessments. They performed femur bone mineral density (FBMD) and ultrasound on the rectus femoris muscle mid-tight cross-sectional area (RFMTCSA) in the quadriceps muscle. The significant changes in BMI were seen in all patients, pre-rehabilitation, BMI = 30 ± 1.06 kg/m^2 and post-rehabilitation, BMI = 29 ± 1.00 kg/m^2. In Pearson's correlation of r = 0.607 between T-score and Z-score in FBMD and RFMTCSA in pre-rehabilitation, there is a little bit significant correlation between the variables than in the Pearson's correlation of r = 0.910 in post-rehabilitation, P < 0.00. Comparing femur bone and rectus femoris muscle parameters as indicators for diagnosis of sarcopenia in chronic respiratory patients, we observed that in rectus femoris muscle, ultrasound is the most effective foot muscle detector.

Keywords: sarcopenia, femoral bone mineral density, rectus femoris muscle phenotype, quadriceps muscles, chronic respiratory disease, quality of life

1. Introduction

Chronic respiratory disease (CRDs) is widespread worldwide. It was reported to be the sixth leading cause of death in the world in 1990 and is now the fourth leading cause of death and is projected to be the third leading cause of death in the world by 2020 [1]. Chronic obstruction of airflow is an important feature of these patients. CRDs have impaired airway function and lung structures. Some of the most common chronic obstructive pulmonary diseases, COPD, asthma, occupational lung disease, pulmonary hypertension and respiratory problems associated with the patient are due to adverse physical conditions such as for overweight and obesity. Patients with chronic respiratory illness can suffer from other symptoms such as frailty, depression, heart attack, fatigue, decreased exercise capacity, and kidney pain. Exercise is in many cases a strategic way to improve the symptoms of these diseases. It is now widely reported that proper exercise can be an effective prevention and treatment strategy for respiratory patients, and this is very important in the management of elderly people with chronic respiratory diseases. Recent studies of extra-pulmonary issues in chronic respiratory patients have shown that quadriceps muscle is the most important muscle for these patients due to involvement in movement and activities [2, 3]. But accurate and reliable equipment must be used to evaluate the quadriceps. New scientific studies have used ultrasound technology as an important and valid device for an accurate evaluation of quadriceps function and structure.

According to the latest scientific reports [4, 5], the rate of sarcopenia in COPD patients is about 25%. Severe sarcopenia was 2.5% and only 0.8% obesity associated with sarcopenia. The incidence of sarcopenia in COPD patients is 15%. It is well known that respiratory disorders are a very common disease. It is also quite evident that respiratory distress is a very common disease that affects up to 10% of adults over 40 years of age and causes high levels of illness and mortality. It is associated with additional respiratory disorders, such as cardiovascular disease, osteoporosis, depression, and anemia. Besides, other non-pulmonary complications, such as cardiovascular disease, bone loss, musculoskeletal disorders, and muscle weakness, can also adversely affect their health outcomes. Increasing different comorbidities can damage lung function, decrease quality of life, and increase mortality. In all of these problems, muscle weakness and osteoporosis is a major problem that needs more therapeutic intervention. In the general population, osteoporosis risk factors include female gender, age, low body weight, glucocorticoid intake in chronic patients, and endocrine problems such as hyperthyroidism and primary hyperparathyroidism. Recently, reduced skeletal muscle as sarcopenia parameter has been identified as a risk factor [6]. The major risk factors for osteoporosis in respiratory patients are not yet clearly understood, but factors such as aging, female gender, low body weight, and body mass index (BMI) have been associated with reduced BMD in patients with COPD [6]. But compared to skeletal muscle index, body weight and BMI do not provide a more accurate reflection of body composition. In one study of aging and body composition, the prevalence of sarcopenia in the overweight group (BMI = 25–29) was 8.9% and in the obese group (BMI > 30) 7.1%. Overweight and obesity are often symptoms of sarcopenia and gradually increase with the prevalence of chronic respiratory disease. Low body mass index is associated with osteoporosis risk factors in patients with COPD including low body weight and low BMD [7]. Sarcopenia is a major complication of chronic obstructive pulmonary disease, which is often seen even without low BMI. However, few studies have been published on the relationship between sarcopenia and BMD.

1.1 Skeletal muscle dysfunction in CRDs

Musculoskeletal disorders are an important clinical examination that is recognized in chronic respiratory patients. For example, in people with COPD, common changes in the musculoskeletal system, including quadriceps weakness, atrophy, and fiber-type shift, each provide independent predictive information of the lung.

The mechanism that disrupts musculoskeletal function can have negative consequences through the progression of the so-called "healthy age" to sarcopenia and weakness. Sarcopenia has been described as a decrease in skeletal muscle and a decrease in physical function dependence, which requires knowledge of current conditions of the musculoskeletal system to reduce muscle mass and muscle weakness in chronic respiratory patients. Skeletal muscle function is often considered in common diagnostic criteria, due to muscle weakness and a history of weight loss, which is often a product of muscle wasting. Both syndromes indicate skeletal muscle dysfunction, which affects more of these syndromes, the broader effects of the disease, both inside and outside the lungs, affecting morbidity and mortality in these patients [8]. Therefore, the presence of sarcopenia may be considered and provide additional prognostic information provided by skeletal muscle function markers. Previous studies have reported that a decrease in skeletal muscle mass is associated with a decrease lung function in patients with COPD. More recently, in studies of nursing home residents, Carlson et al. [6] showed that peripheral muscle strength, including hand-grip strength, was associated with maximal stimulating muscle strength. However, there is a paucity of information on how skeletal muscle mass changes are associated with pulmonary function in adults without lung disease.

1.2 Recognizing sarcopenia in CRDs

Some scientific reports suggest that one of the systemic effects of COPD is sarcopenia. The term is described as an age-related decline in muscle volume and function. This situation is associated with negative health consequences such as falling, disability, hospitalization, poor quality of life and mortality. The cause of sarcopenia is in addition to the consequences of the disease, nutrition, and activities caused by physiological changes. Sarcopenia can be classified as a physical impairment associated with adverse health consequences. These findings suggest that sarcopenia is associated with decreased lung function in COPD patients. These patients also have a relative or absolute increase in fat mass, which can lead to systemic inflammation, and insulin resistance [9, 14]. In a recent study, body mass index (BMI) was not significantly associated with lung function, the severity of dyspnea, quality of life, and decreased skeletal muscle mass.

Skeletal muscle dysfunction is a well-known clinical manifestation in COPD patients. Key features include quadriceps weakness, atrophy, and type II fiber alteration, all of which are associated with a poor prognosis independent of lung function. Sarcopenia describes age-related skeletal muscle loss, leading to an increased risk of physical disability, poor health, and mortality. Sarcopenia is increasingly recognized as a clinical syndrome with its contributing factors, including physical inactivity, malnutrition, and chronic illness. Since COPD is in some ways an accelerated disease in the aging process, it is proposed the hypothesis that sarcopenia is related to COPD patients [10]. In patients with COPD, most studies addressing skeletal muscle dysfunction have focused on one aspect of sarcopenia, mainly in the lower limbs. This contradicts international consensus statements on sarcopenia, which emphasizes the loss of both muscle mass and function in diagnostic criteria, and emphasizes the importance of general muscle function. In particular, evaluation of one aspect of sarcopenia is not sufficient, as the relationship between muscle mass and strength is nonlinear, and muscle atrophy does not always lead to dysfunction and there are no functional status weaknesses. The European Working Group on Sarcopenia in the Elderly (EWGSOP) has developed practical clinical diagnostic criteria for sarcopenia approved by international organizations and used to assess the prevalence and impact of this syndrome on disease settings and states. Although commonly used in COPD, it is necessary to understand the magnitude and nature of the problem in the disease associated with atrophy and muscle weakness [11]. Sarcopenia is associated with many common disease management strategies, including exercise training and nutritional aspects. Given the emergence of drugs directed to sarcopenia in other disease conditions, such data may be useful for drug production [12]. In this study, we evaluated the prevalence and risk factors of sarcopenia in respiratory patients and the effect of sarcopenia on functional exercise capacity and health status. We also seek to examine the relationship between sarcopenia and quadriceps strength and to examine whether exercise training as part of pulmonary rehabilitation can reverse sarcopenia.

Patients with respiratory disorders have musculoskeletal disorders, but their function increases with the progression of physical activity and reduces the risk of general and muscular weakness. Respiratory patients with general and muscular weakness have higher mortality rates than non-weak patients and are more likely to have sarcopenia and an increased incidence of the disease. In one study at a British hospital (2015) [13], the prevalence of sarcopenia was reported to be 14.5% of COPD patients in British country compared to other European countries. In chronic respiratory patients, both risk factors (smoking, aging) and their causal mechanisms (endocrine dysfunction and inflammatory cytokines) are common and be high prevalence. These causes have increased with age and the global prevalence of recurrent respiratory diseases. Muscle structure and function must be considered for the diagnosis of sarcopenia. Clinically, the current definition of sarcopenia may show several defects, especially for quantitative measurement of muscle volume. Firstly, muscle mass thresholds are defined differently, and this causes patients to be classified correctly or incorrectly for sarcopenia. The prevalence of sarcopenia in the elderly also depends on the accepted definition for evaluation. However, the role of skeletal muscle ultrasound for screening and diagnosis of sarcopenia in the elderly remains and is important. None of the current definitions of sarcopenia include it in the diagnostic algorithms currently in the category of specialists. However, some experts [14–16] believe that using ultrasound in this field is also useful and this technique is recognizable based on pioneering studies of muscle mass in healthy individuals and patients with an aging approach. Therefore, most studies support the use of potentially validated muscle ultrasound to identify sarcopenia in the elderly. However, since it has been performed in small samples and a variety of clinical conditions (from healthy subjects to patients with chronic diseases), no significant recommendations have been made regarding the use of large-scale muscle ultrasound. In smaller cases, so in the same patients, some muscles may be affected by sarcopenia and other muscles not affected. Innovative muscle ultrasound studies have been conducted by Abe et al. [19] and his colleagues have contributed to the development of knowledge of this phenomenon and have developed specific concepts. They also developed and validated the equations, and calculated the total body mass index from ultrasound muscle thickness measurements in Japanese and Caucasian subjects, and achieved significant results. However, the relationship between full-body sarcopenia and specific sarcopenia is not fully understood and needs further research to identify the indicators. In this capture, in a small group of healthy adults, the researchers showed that the ratio of anterior or posterior muscle to ultrasound was not consistent with abdominal lumbar mass measured by DEXA. However, according to the researchers' findings, it can be concluded that using ultrasound and DEXA to predict sarcopenia indices in chronic respiratory patients is valid and reliable, but which parameter has the most impact? It is not clear yet and we need to investigate more in the future. However, the role of skeletal muscle ultrasound for screening and diagnosis of sarcopenia in the elderly is quite clear. But none of the current indicators of sarcopenia include its diagnostic algorithm. Bonsignore [17] and Marmorato et al. [18] believe that the use of muscle ultrasound is also potential in this area and is largely based on pioneering studies in which muscle mass and its architecture are evaluated using this method in healthy subjects and respiratory patients.

1.3 Sarcopenia indexes in CRDs

1.3.1 Body mass index (BMI)

Body mass index (BMI) should be mounted on a wall using standard hospital calibration scales, because these parameters are very useful and fruitful in respiratory treatment as an indicator of health status. BMI is calculated as body mass (kg) divided by squared body height (m^2). However, this mostly applies to patients with severe COPD where an increasing BMI is linearly associated with better survival, while in patients with mild to moderate COPD the lowest mortality risk occurs in normal to overweight or weight loss in these patients. The World Health Organization criteria were used to classify the subjects as low-weight (BMI < 18.5), eutrophic (18.5 < BMI ≤ 24.99), overweight (25 ≤ BMI ≤ 29.99) or obese (BMI ≥ 30.00) [19]. This index and division for respiratory patients can also be cited and so for all population recognized. The biggest problem with BMI is that when patients with chronic respiratory disease have a normal weight, they are unable to recognize the percentage of the muscles of this patient, which is the main cause of her movement and activity, and in this situation, we need to use more precise equipment the body composition by BIA (Bioelectrical Impedance Analysis), including the percentage of muscle, fat, bone, lean mass, hole body mass, is more accurately measured, and this can be useful in a more accurate diagnosis of sarcopenia in these patients [20].

1.3.2 Skeletal muscle index (SMI)

One of the most important indicators of sarcopenia in chronic respiratory patients is the skeletal muscle index (SMI), which is measured by factors such as age, height, weight, ethnicity, gender, and BMI in a valid and reliable formula, and its rate in an evaluation table according to gender and age is measurable. This index is the most important factor in the diagnosis of sarcopenia in chronic patients, especially in patients with respiratory disease. SMI is calculated as a function of weight and height as follows: (height [m]_0.244_body mass) + (7.8_height) + (6.6_gender) − (0.098_age) + (ethnicity − 3.3). The SMI index is then calculated by dividing an individual's SMI (kg) by his or her height squared (m^2). This indicator can be used as the main factor in the diagnosis of a respiratory patient with sarcopenia [21].

The gold standard of research for evaluating sarcopenia relies on complete techniques, cross-sectional imaging, a non-functional, and more structured approach to routine care. A more practical alternative indicator of lumbar muscle density in L3 using a normalized computed tomography (CT) is called skeletal muscle index (SMI). While evidence suggests that decreased lumbar SMI is associated with adverse clinical outcomes, such as deaths in the lung or colorectal cancers, little research has investigated how this measure of sarcopenia relates to dyspnea or decreased exercise tolerance. Although lung cancer patients and respiratory patients often use chest CT scans as part of their care, fewer respiratory patients receive lumbar scans. This limits the ability to evaluate sarcopenia using the lumbar SMI and therefore requires the discovery of the quadriceps SMI instrument as a more accurate sarcopenia measure. Besides, the measurement of thoracic skeletal muscle, which is involved in breathing work, may be associated with better breathing and better functional capacity [22, 23].

Spencer [24] and Soicher et al. [25] did not report a significant relationship between SMI and breath intensity. There was also no significant relationship between SMI, respiratory rate, and 6MWT interval. Similarly, the Cox proportional hazards model [25] did not show a significant relationship between SMI and manual weakness. Finally, using this technique, 50 patients with eligible Lumbar Scan diagnoses were identified and found similar results. Over time, the SMI has gradually declined unacceptably. There was a significant relationship between Pearson correlation coefficients in lumbar and thoracic scans in this issue. Their findings suggest that the definition of SMI-based sarcopenia is not associated with severe breathing, exercise capacity, or survival in a small sample of patients with advanced lung cancer. The strengths of the present study include a population with complementary sarcopenia features, severe breathing and exercise tolerance, and robust exploratory analysis. Despite the negative results, they demonstrated the feasibility of measuring sarcopenia using SMI. They were limited by the small sample size and missing data. Whereas a larger sample provides more power to detect significant index correlations. They used CT scans performed in the usual stages of care, which may not meet the exact criteria of future research. Changes in the quality of CT scans may result in indeterminacy. Besides, it cannot illuminate the severity of respiratory illnesses and other complications present in diagnostic models to potentially improve the accuracy of treatment models in respiratory patients [26, 27].

1.3.3 Anthropometric indexes

For a more accurate diagnosis of sarcopenia, according to scientific reports, anthropometry and measurement of body sections such as arms, trunk, pelvis, and legs are important parameters for measuring anthropometry in the diagnosis of sarcopenia. A chronic respiratory patient must be normal, since muscle atrophy will be directly related to muscle weakness and general weakness of the body, and ultimately lead to a decrease in the physical activity of the patients. In these conditions, the quality of life of the patients is compromised and they are not able to continue their normal life and eventually the mortality rate increases [28].

1.4 Pulmonary rehabilitation in CRDs

The results of scientific predictions show that pulmonary rehabilitation (PR) reduces frailty but there is little evidence of this intervention in this area. PR has been shown to significantly improve patients' symptoms and quality of life in patients with respiratory disease. Public daily activities can relieve shortness of breath and fatigue, as well as increase exercise tolerance, and affect patients' self-control and feel it. Recently, studies have shown that the lack of association between fat mass and the 6-minute walk test (6MWT) is one of the general considerations in assessing COPD patient status longitudinally to identify alternatives in predicting the future of these patients. In addition to improving respiratory and functional symptoms, pulmonary rehabilitation programs also target elements such as weakness, depression, inactivity, and fatigue [29, 30].

Jones et al. [31] investigated the interaction of sarcopenia index in patients with COPD and response to pulmonary rehabilitation. In this study, 622 elderlies and middle-aged COPD patients were included in the study. An immediate cohort study was followed over four years of pulmonary rehabilitation in patients with weakness and COPD. The pulmonary rehabilitation program consisted of an 8-week outpatient and 2-time weekly and home-based one-time training program.

The sessions consisted of 1:15 h of training, with 25.6% of participants in the pulmonary rehabilitation program being a weakness (according to the Freud phenotype model), while only 10% of the participants did not meet any of the weakness criteria. Significant improvements have been reported in a variety of areas including the Dyspnea MRC scale, manual dynamometer, chronic fatigue and anxiety, emotional scores, hospital stress, and depression shuttle score and walking test. All of these parameters are related to sarcopenia indices in respiratory patients. Sarcopenia has increased with age and the World Initiative Index for Obstructive Pulmonary Disease. It can be clearly stated that disorders of the skeletal muscle are more important in evaluating sarcopenia in chronic respiratory patients [32, 33]. In the event of any disruption to the structure and function of the large musculoskeletal system of the body, especially the lower limbs and the foot, which are the main cause of movements, there will be widespread changes in weight loss, overweight, body composition, body diameter, water, fat, muscle percentage. Ultimately, the amount of physical activity a patient has directly related to their muscles [34, 35].

2. Main objective

The main objective of this study was to evaluate the relationship between rectus femoris phenotype and femoral bone mineral density as the main indicators of sarcopenia in chronic respiratory patients following a pulmonary rehabilitation protocol with a cardiopulmonary exercise test approach (ERS/ATS instructions). The effect of this relationship on lung function and muscle structure in these patients has also been investigated.

3. Methodology and materials

3.1 Ethics and Research Committees

This study has been approved by the Ethic Committee and the Research Committee Parc Sanitari Sant Joan de Deu, in the research group: "Clinical and epidemiologic research on high-prevalence disorders" (Faculty of Medicine at the University of Barcelona). All patients who volunteered before signing an informed consent form had all information about the goals, techniques, possible outcomes, and therapeutic processes in the pulmonology, rehabilitation, and radio diagnostics departments. Also, all patient information without personal access is completely confidential and is for the sole purpose of this research (**Figure 1**).

3.2 Study design

Figure 2 shows the general design of the study as a work plan of the study. The patients had asthma, COPD, bronchiectasis and obesity (randomly, 38 men and women selected from the chronic respiratory community at the hospital) with a dyspnea score ≥ 2 in MRC index. We evaluated the general characteristics of the patients including: gender, age, weight, height, BMI and clinical history, spirometry of lung function, at the pulmonology and rehabilitation departments. Then, they were referred to the Radio diagnostic department to perform a DEXA scan test to evaluate the femoral bone mineral density (T-score and Z-score) and ultrasonography on the rectus femoris quadriceps muscle (cross-sectional area, distance and circumference). They performed a 4-month long term a pulmonary rehabilitation protocol (**Table 1**), which included: exercise tests—incremental and constant,

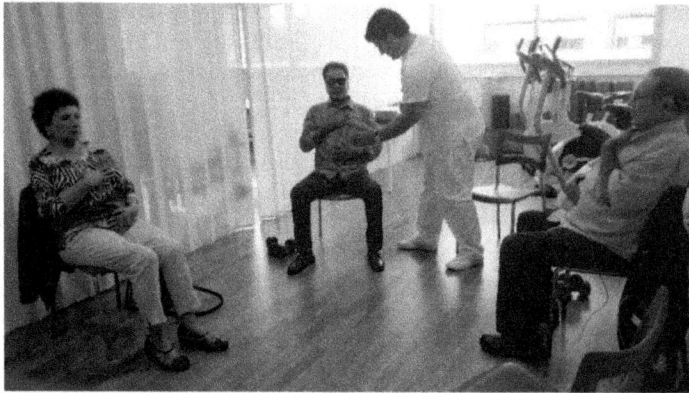

Figure 1.
Group sessions of chronic respiratory patients in Department of Rehabilitation at Parc Sanitari Sant Joan de Deu.

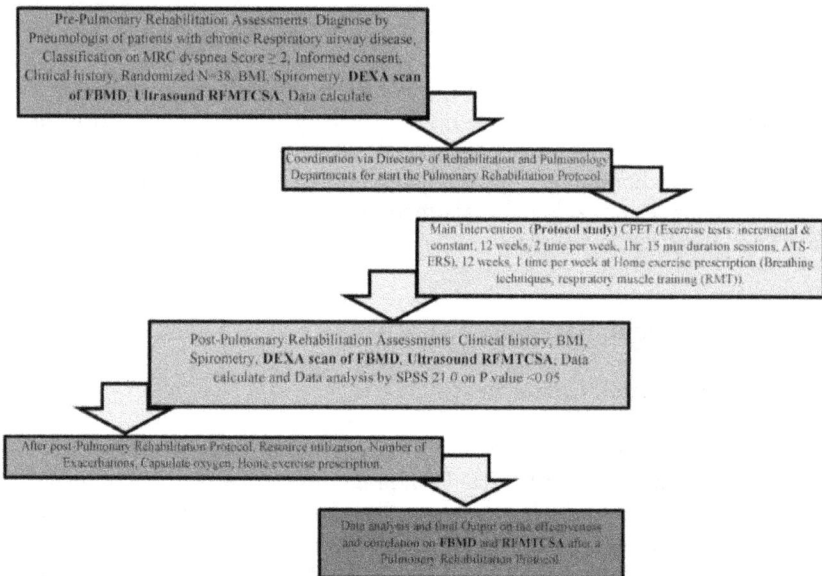

Figure 2.
Work plan concept (pulmonary rehabilitation protocol).

12 weeks, 3 times a week, duration of each session was 1 h 15 min. Breathing techniques, respiratory muscle training and self-management, (ATS-ERS guidelines, 2013–2015) were conducted with the supervision of specialists. Following the pulmonary rehabilitation protocol, all patients performed assessments as post-rehabilitation like pre-rehabilitation, according to the protocol study. *Main intervention*: cardiopulmonary exercise tests (CPET) was developed by American Thoracic Society in 2003 [36] as the gold standard with validity and reliability to study a patient with limited level of exercise and to evaluate improvement of respiratory patients before and after pulmonary rehabilitation protocol.

3.3 Statistics

Descriptive statistics (frequency, mean values ± standard deviation, variance...) was used to analyze the variables of clinical history and health status indexes in

No	Exercise	Instructions	Time (min)
1	Warm up	Walking, rotation of joints in upper limbs and lower limbs, rotation and stretching of trunk in low back, chest, neck and shoulders and quadriceps	15
2	Cycling ergometer or treadmill	The test began with a 1-min warm-up period at minimal cycle ergometer load (15 W), with 5- to 10-W increases every 2 min that were individually selected to maintain the period of load increase in the 8-to-12-min range. 1-min active recovery using minimal cycle ergometer load followed the peak load interruption and was followed by a 6-min passive recovery. Treadmill test is special for patients that they have knee osteoarthritis and must be attention to standards indexes in exercise program and initially, the walking speed is very slow for warm up, but each minute the required walking speed progressively increases. Total, time for treadmill test 6 min	25
3	Light dumbbell	Repetition of light dumbbell (50% resistance) for improvement of endurance muscle in major muscles, e.g., shoulders, back, low back, pectoral muscles, trunk sides, quadriceps, leg muscles that more used in exercise program and influence on breathing	10
4	Respiratory muscle training	RMT may consist of inspiratory muscle training (IMT) or expiratory muscle training (EMT) or a combination of both includes: (1) diaphragmatic reeducation, (2) profound inspiration, (3) inspiratory hiccups, (4) resistive inspiration with linear pressure load	10
5	Breathing techniques	(1) Pursed-lips breathing, (2) diaphragmatic (abdominal/belly) breathing, (3) better breathing tip: stop, reset, continue	5
6	Cold down	Light walking, deep breathing, stretching of muscles that more used in exercise program, e.g., breathing muscle, major muscles, peripheral muscles that employed during exercise and fresh mind 1 min	10

Table 1.
Cardio-pulmonary exercise protocol (study protocol) in chronic respiratory patients.

preliminary characteristics (age, gender, height, weight, BMI, patients, smoking, diabetic, hypertension, depression and crisis).

Differences between pre- and post-rehabilitation protocol in assessments of spirometry, skeletal muscle index (SMI), femur bone mineral density and ultrasound rectus femoris muscle in quadriceps, were analysed using t-Student of one sample T-test, independent sample T-test and paired sample T-test. Evaluation of correlation between rectus femoris phenotype parameters and femur bone mineral density indexes was done using Pearson correlation in bivariate method of SPSS.

In this study we received report analyzed from department of electrodiagnostic on rectus femoris muscle by ultrasound report and femur bone mineral density by DEXA scan that we used of these figures in assessments of rectus femoris phenotype and FBMD parameters.

We evaluated all of the variables analysis by SPSS version 21.0 software (SPSS Inc., 2012, Chicago, IL, USA) and Excel 2016 (office 2016) to be used for database. Statistical significance was set at $P < 0.05$.

4. Results

4.1 Measurement preliminary characteristics and clinical history

Table 2 shows that, most of the patients had COPD and the prevalence was higher in men. We found significant changes in chronic respiratory patient's BMI

Variables		Frequency	Valid Present (%)	Mean ± SEM	SD	Variance	Minimum	Maximum	Percentile100
Gender	Male	22	57						
	Female	11	28						
Missing		5	13						
Age (yr)				72 ± 1.0	7.0	54.0	53	88	76.0
Patients	COPD Men	18	46						
	Women	4	10						
	Asthma Men	1	2						
	Women	5	13						
	Bronchiectasis Men	1	2						
	Women	2	5						
	Obesity Men	2	5						
	Women	0	0						
BMI—pretest (kg/m²)				30 ± 1.06	6.0	37.0	17	45	33.0%
BMI—posttest (kg/m²)				29 ± 1.00	5.0	33.0	17	40	33.0%
Height (cm)				166 ± 1.0	10.0	108.0	146	196	172.0%
Weight—pretest (kg)				85 ± 3.0	21.00	471.0	45	142	95.0%
Weight—posttest (kg)				84 ± 1.0	21.04	442.0	45	141	94.0%
Recent Hospitalization	Yes	16	42						
	No	17	44						
Smoking (yr)				15 ± 3.0	17.00	323.0	00	60	25.0
Capsulate oxygen (h/day)	Pre-RHB	33	86	2 ± 0.0	1.0	1.0	1	6	3.0
	Post-RHB	33	86	2 ± 0.0	0.0	0.0	0	4	3.0

Variables		Frequency	Valid Present (%)	Mean ± SEM	SD	Variance	Minimum	Maximum	Percentile100
Arthritis	Yes	8	50						
	No	25	36						
Cardiovascular disease	Yes	12	21.1						
	No	21	65						
Depression	Yes	12	31						
	No	21	55						
Diabetic	Yes	14	31						
	No	19	55						
Hypertension	Yes	High	36						
	No	Middle	50						
Crisis	Pre-RHB	High	14	36	Post-RHB	High	3	7%	
	Middle	3	7		Middle	10	26%		
	Low	10	26		Low	20	52%		

Table 2.
Preliminary characteristics and clinical history pre- and post-rehabilitation protocol at the study on P < 0.05.

from 30 ± 1.06 (kg/m^2) before the rehabilitation protocol to 29 ± 1.00 (kg/m^2) after that and we also observed weight loss in all respiratory patients. Regarding the clinical history of patients, we can see that 42% of patients have a history of hospitalization and the average rate of smoking in all of them is 15 ± 3.0 years. A remarkable point in this study, which is very important for respiratory patients, is the amount of oxygen consumed via O_2 capsulate at home which significantly decreased special in COPD and obesity patients (from 30% to 14%) after the pulmonary rehabilitation protocol (**Figure 3**). **Table 2** also shows that 50% of the patients had knee arthritis, 21% cardiovascular disease, 31% depression, 31% diabetes, and 14% hypertension. Finally, the rate of respiratory crisis with high intensity decreased significantly ($P < 0.05$) from 14% to 3% after the pulmonary rehabilitation. We observed that in all clinical and general health factors, chronic respiratory patients recovered, presenting improvement in general health and clinical conditions.

4.2 Measurement of correlation between FBMD and RFMTCSA

In the analysis of Pearson's correlation r = 0.607 between T-scores and Z-score in femur bone mineral density (FBMD) and Rectus femoris Mid-Tight Cross Sectional Area (RFMTCSA) in pre-rehabilitation, there is a significant correlation between the variables ($P < 0.001$). In the analysis of the Pearson's correlation r = 0.910 in post-rehabilitation between T-score and Z-scores in FBMD and RFMTCSA, there have a higher significant correlation between variables than pre-rehabilitation on $P < 0.001$ (**Table 3**). It can be said that there is a significant relationship between FBMD and RFMTCSA after pulmonary rehabilitation protocol, and in both T-score and Z-score of FBMD with rectus femoris phenotype in RFMTCSA significant

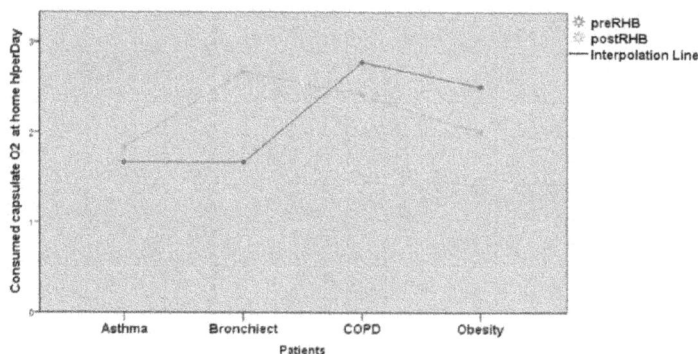

Figure 3.
Oxygen capsulate consumption monitored at home, h/per day, P < 0.05.

Variables	One-Sample Test				
	M	df	P	Lower	Upper
Skeletal muscle index pre-RHB	16 ± 32	32	0.001	28.00	36.00
Skeletal muscle index post-RHB	17 ± 34	32	0.001	30.00	38.00

*Formula: *Skeletal muscle index (SMI): (height [m] 0.244 body mass) + (7.8 height) + (6.6 sex) − (0.098 age) + (ethnicity − 3.3).*

Table 3.
T-test skeletal muscle index on chronic respiratory airway patients on P < 0.05.

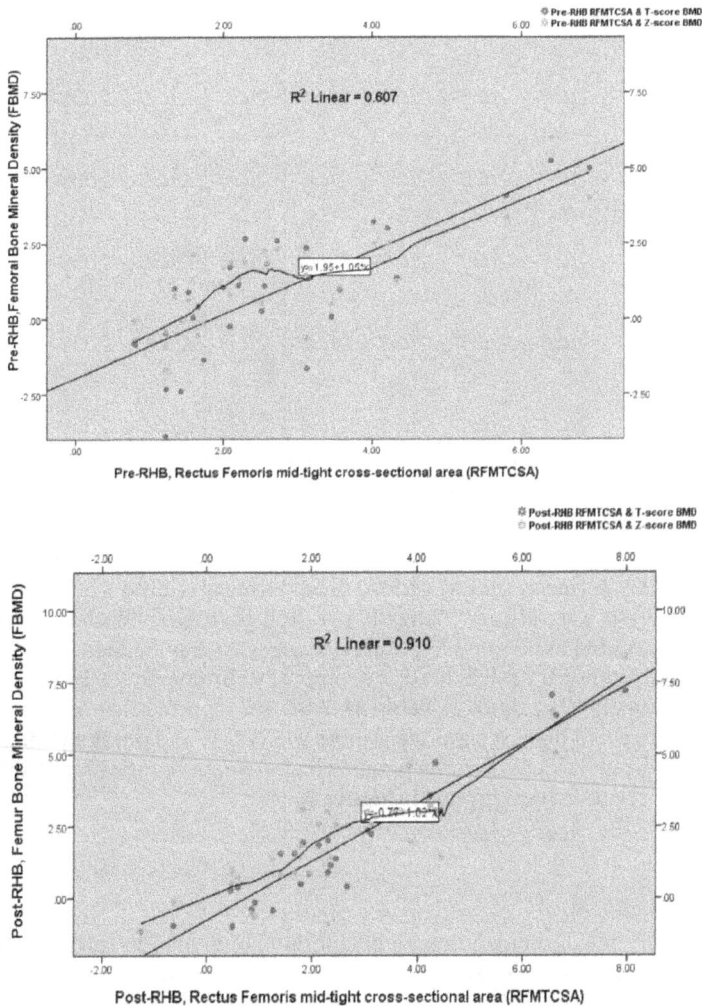

Figure 4.
Correlation between T-score and Z-score of the femur bone mineral density and rectus femoris cross sectional area in phenotype quadriceps, at P < 0.001.

progress was observed. But in chronic respiratory patients this significant increase was shown in Z-score higher than T-score. Also we can see that in circumference of RFMTCSA between pre- and post-rehabilitation protocol did not have significant change or improvement in correlation with T-score and Z-score (**Figure 4**).

4.3 Measurement of skeletal muscle index (SMI)

There was a significant increase in musculoskeletal index before (pre-rehabilitation SMI, mean = 16 ± 32) and (post-rehabilitation SMI, mean = 17 ± 34) after the pulmonary rehabilitation protocol and they showed positive changes after the rehabilitation protocol $P < 0.001$ (**Table 3**).

4.4 Measurement of spirometry lung function

In the spirometry variables, we can see that most significant changes have occurred in FEV1/FVC % (mean $\pm$ SEM = 40.00 ± 7.0 vs mean $\pm$ SEM = 35.00 ± 7.0)

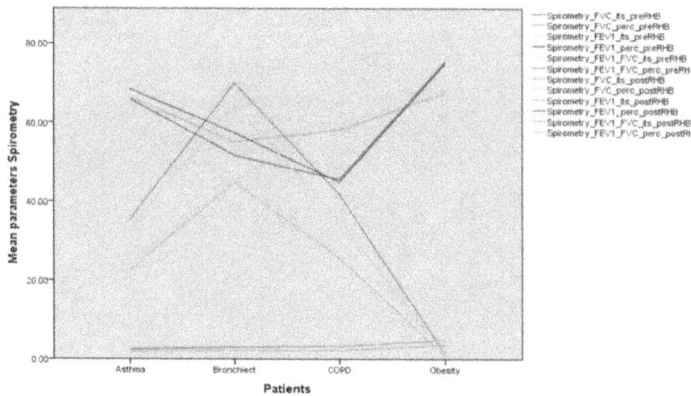

Figure 5.
Mean different between spirometry parameters (liter, %) in asthma, bronchiectasis, COPD and obesity patients in pre- and post-pulmonary rehabilitation protocol.

and FEV1/FVC ltr (mean ± SEM = 27.00 ± 5.0 vs mean ± SEM = 25.00 ± 5.0). In variable of FEV1 % (mean ± SEM = 51.00 ± 2.0 vs mean ± SEM = 52.00 ± 2.0), we saw that there was a significant change in pre- and post-rehabilitation protocol that cannot be considered as acceptable positive changes but can be seen in the effectiveness of the pulmonary rehabilitation protocol on spirometry variables in chronic respiratory patients (P < 0.001) (**Table 3**). Also, we reported that in bronchitis patients we have had highest significant in FEV1/FVC % and obesity patients had more significant in FEV1 %. Asthma patients had increase significant in FVC perc in pre- and post-rehabilitation protocol (**Figure 5**).

5. Discussion

In this study, we concluded that by comparing bone, muscles as the main indicators diagnosis of sarcopenia in chronic respiratory patients, we find that rectus femoris is the most effective quadriceps muscles in the diagnosis of sarcopenia. RFMTCSA, Distance, and Circumference muscle parameters, and DEXA scan at Z-score and T-score of femur bone density after patients' rehabilitation protocol were significantly different. In other words, in all the ultrasound indices of the rectus femoris muscle and the DEXA scan data of the femur, we found positive incremental data. With these considerations in mind, it can be said that although patients with normal daily activities can control the main indicators of sarcopenia, they can improve the quality of life and improve the outcomes of all major muscular and bone markers. Muscle function, lung function, and exercise capacity require a comprehensive pulmonary rehabilitation program that can help to prevent muscle weakness and degeneration, as well as reduce important bone mineral density such as the femur and ultimately their mortality rate. We conclude that not all factors are necessary to accurately determine the severity of sarcopenia in patients with chronic respiratory disease, and if specialists consider rectus femoris ultrasound and femur bone DEXA as the most important factors for maintaining body stability, they can monitor the latest status of respiratory patients with sarcopenia and more accurately diagnose the health status and mortality of the chronic respiratory patients.

We also examined whether spirometry indices are actual measurements and better predictors of whether CRDs patients develop sarcopenia than the GOLD staging of the disease. Sarcopenia and obesity have direct effects on lung function values and we recommend that both elements should be evaluated in all respiratory patients. We found that respiratory patients with sarcopenia had lower T-score BMD, which means a higher prevalence of osteoporosis. Also, the presence of sarcopenia was significantly associated with an increased risk of osteoporosis and low BMD in both low- and high-weight patient groups. Therefore, in evaluating patients with CRDs, the presence of sarcopenia should be considered an independent risk factor for low BMD.

Lung function is a stronger independent predictor compared to body composition parameters. Taken together, these parameters are not only useful for assessing breathing and the treatment process, but are also a potential surrogate for patients' physical function. Some of the limitations of our study deserve attention. The relatively small size of our samples, which may be in line with the study hypothesis, may limit our findings. But our conclusion holds for a whole population, but may not be entirely true for a population with more prevalent respiratory disease. This does not mean that the conclusion is highly generalizable. We evaluated body composition by DEXA, a method widely accepted in scientific research, but not by the "gold standard" for patients. Besides, we define sarcopenia based on purely quantitative parameters of body composition, but we are also able to provide information on the qualitative dimension of sarcopenia. Finally, the cross-sectional design of the study allowed for the creation of a causal sequence among the relationships studied. Body composition may be an important determinant of physical function in elderly respiratory patients. This research is needed to confirm these observations on larger samples to improve our understanding of the factors influencing physical function in a more complex population such as chronic respiratory patients.

In a study similar to our study, Xiong et al. [13] mentioned the intensity of echo (EI) was significantly increased by ultrasonography in rectus femoris at all stages of the GOLD standard following the pulmonary rehabilitation protocol for COPD patients. However, the cross-sectional area of rectus femoris (RFcsa) in GOLD standard decreased. This suggests that changes in EI Rectus femoris may occur earlier in muscle size in patients with COPD. EI was not associated with age, BMI, and airflow obstruction in our patients. These results show peripheral muscle function in patients with COPD, whose strength and size of the quadriceps are lower than in normal subjects. Some researchers have focused on the relationship between quadriceps strength and thickness with lung function in respiratory patients, which can be assessed using muscle ultrasound [37]. In another study, the effect of small and large muscular disorders was examined on admission, especially in quadriceps.

Rongchang [9], reported expected gender differences in quadriceps muscle for COPD and control groups and similar data were also reported, indicating that in quantifying quadriceps performance compared to controls, sex and age, mean QMVC declined 47% and 45% in men and women; a more severe decrease than previously reported, indicating a significant impairment of quadriceps strength in COPD patients. In other studies, the effects of small and large muscular disorders, especially in the quadriceps, was related to the time of hospitalization.

In the study Greening et al. [4] measured the level of quadriceps (Qcsa) as a marker of muscle mass. This method has previously been confirmed in computed tomography, Qcsa as well as X-ray absorptiometry in people with COPD and is useful for patient evaluation in acute conditions and under the need of patient effort. Regarding the hospital admission, the mean daily proportion of patients with respiratory distress and disorder was different in the small and large muscle groups. This indicates that skeletal muscle function is crucial not only for hospital admission but also for the severity and duration of hospitalization.

5.1 Response to pulmonary rehabilitation (PR)

As previously mentioned, in patients with chronic respiratory airway disease, sarcopenia is one of the major problems of these patients, which gradually causes atrophy and weakness in their muscles, ultimately reducing physical activity and severely affecting their quality of life. If they do not participate in pulmonary rehabilitation and physical exercise, they will experience general weakness and a gradual increase in mortality. Therefore, it can be said that the major part of sarcopenia is due to musculoskeletal disorders, which include their structure and physical movement, and should be a top priority in these patients' pulmonary rehabilitation programs. In this study, we aimed to investigate sarcopenia factors in chronic respiratory patients that revealed a significant increase in ultrasound on the rectus femoris cross-sectional area (RFMTCSA) and rectus femoris peripheral as well as on DEXA scans on femoral bone mineral density (FBMD) at the Z-score, in fact, rectus femoris muscle peripheral and femur bone density were among the factors that had positive effects on factors such as leg muscle strength, quality of life, lung function and exercise capacity. Regular rehabilitation programs have shown that they have progressed gradually, but their sarcopenia has dropped significantly, requiring regular, long-term pulmonary rehabilitation.

6. Conclusions

6.1 Our programme of pulmonary rehabilitation improved

- Parameters of quadriceps muscle specially in COPD patients, RFMTCSA in pre-rehabilitation = 1.73 and post-rehabilitation = 2.38.

- The rate of femur bone mineral density (FBMD) increased after the pulmonary rehabilitation protocol.

- We found that there was a significant reduction in BMI and crisis indicators in respiratory patients, and there was significant difference in the low rate consumption of oxygen capsule.

- Correlation between RFMTCSA quadriceps and FBMD factors significantly increased after the pulmonary rehabilitation protocol.

- Significant positive changes were observed in FEV1/FVC *its*, FEV1/FVC *perc*, FVC *its*, we have had same record scales before and after the pulmonary rehabilitation protocol.

- Also, patients were also less likely to use capsulate oxygen at home and we found that home exercises and breathing techniques had a significant impact on maintaining the effects of the pulmonary rehabilitation protocol and enhancing self-esteem and confidence in the daily activities of the respiratory patients, and to the extent that they did so long after the rehabilitation protocol, exercises, and breathing techniques continued at home.

I hope this research will shed some light on the complications of sarcopenia in chronic respiratory patients and help to reduce the problems of older patients with chronic respiratory disease in the near future.

Abbreviations

ATS	American Thorax Society
AT	acid lactic threshold
BMI	body mass index
BIA	bioelectrical impedance analysis
BODE Index	body mass index, airflow obstruction, dyspnea and exercise capacity index
CRDs	chronic respiratory disease
COPD	chronic obstructive pulmonary disease
DEXA	dual-energy X-ray absorptiometry
ERS	European Respiratory Society
EI	echo intensity
EWGSOP	European Working Group on Sarcopenia in the Elderly
EMT	expiratory muscle training
FBMD	femur bone mineral density
FEV1	forced expiratory volume per 1 second
FVC	forced vital capacity
FEV1/FVC	expire in the first second of forced expiration (FEV1) to the full, forced vital capacity (FVC)
GH	general health
GOLD	global initiative for chronic obstructive lung disease
HRQL	health-related quality of life
ILT	incremental load test
IMT	inspiratory muscle training
kg	kilogram
L	liter
MRC	Medical Research Council
MVV	maximal voluntary ventilation
MH	mental health
O_2	oxygen
PR	pulmonary rehabilitation
PSSJD	Parc Sanitari Sant Joan de Deu
PF	physical functioning
PI value	protease inhibitor value
pre-RHB	pre-rehabilitation
post-RHB	post-rehabilitation
PTH	parathyroid hormone
RFMTCSA	Rectus Femoris Mid-Tight Cross Sectional Area
RFcsa	rectus femoris cross sectional area
RP	role physical
RE	role-emotional
RMT	respiratory muscle training
SMI	skeletal muscle index
6MWT	6-minute walking test
SpO_2	peripheral capillary oxygen saturation
SF	social functioning
SPSS	statistical package for the social sciences
SAP	intranet PSSJD
US	ultrasound
VO_2	oxygen uptake
VE	ventilation

VE/VCO2	minute ventilation-to-carbon dioxide output
VD/VT	dead space over tidal volume
VC	vital capacity
VT	vitality
WT	walking test
WHO	World Health Organization

Author details

Esmaeil Alibakhshi[1,2,4*], Raffaele Fiorillo[1], Luis Lores Obradors[2],
Ana María Sánchez-Laforga[3], Amparo Villar Cánovas[3], Mostafa Ghanei[4] and
Ali Qazvini[4]

1 Department of Rehabilitation, Parc Sanitari Sant Joan de Deu (PSSJD),
Faculty of Medicine, Barcelona University, Barcelona, Spain

2 Department of Pulmonology, Parc Sanitari Sant Joan de Deu (PSSJD),
Faculty of Medicine, Barcelona University, Barcelona, Spain

3 Department of Electro Diagnostic, Parc Sanitari Sant Joan de Deu (PSSJD),
Faculty of Medicine, Barcelona University, Barcelona, Spain

4 Chemical Injuries Research Center, Systems Biology and Poisoning Institute,
Baqiyatallah University of Medical Sciences, Tehran, Iran

*Address all correspondence to: ealibaal7@alumnes.ub.edu

IntechOpen

References

[1] Franssen FME, Rochester CL. Comorbidities in patients with COPD and pulmonary rehabilitation: Do they matter? European Respiratory Review. 2014;**23**:131-141

[2] Alibakhshi E, Lores L, Fiorillo R. Physiological factors relevant to exercise tests in pulmonary rehabilitation of COPD patients. Journal of Sports Medicine and Doping Studies. 2015;**6**: 2161

[3] Alibakhshi E, Lores Obradors L, Fiorillo R, Ghaneii M, Qazvini A. Gender disparity of changes in heart rate during the six-minute walk test among patients with chronic obstructive airway disease. Australasian Medical Journal. 2017;**10**(7):637-644

[4] Greening TC, Harvey D, Chaplin EJ, Vincent EE, Morgan MD, Singh SJ, et al. Bedside assessment of quadriceps muscle by ultrasound after admission for acute exacerbations of chronic respiratory disease. American Journal of Respiratory and Critical Care Medicine. 2015;**192**(7):810-816

[5] Alibakhshi E, Obradors LL, Fiorillo R, Ghanei M, Panahi Y. Effectiveness of pulmonary rehabilitation on malignant respiratory diseases. Journal of Cardiopulmonary Rehabilitation. 2017;**1**(2):114

[6] Charlson ME, Pompei P, Ales KL, MacKenzie CR. A new method of classifying prognostic comorbidity in longitudinal studies: Development and validation. Journal of Chronic Diseases. 1987;**40**(5):373-383

[7] Maltais F, Decramer M, Casaburi R, Barreiro E, Burelle Y, Debigare R, et al. An official American Thoracic Society/European Respiratory Society statement: Update on limb muscle dysfunction in chronic obstructive pulmonary disease. American Journal of Respiratory and Critical Care Medicine. 2014;**189**(9):e15-e62

[8] Huppmann P, Sczepanski B. Effects of inpatient pulmonary rehabilitation in patients with interstitial lung disease. The European Respiratory Journal. 2013;**42**:444-453

[9] Rongchang CC. Factors associated with impairment of quadriceps muscle function in Chinese patients with chronic obstructive pulmonary disease. PLoS ONE. 2014;**9**(2):e84167

[10] Hakamy A, McKeever TM, Jack E, Gibson E, Bolton CE. The recording and characteristics of pulmonary rehabilitation in patients with COPD using The Health Information Network (THIN) primary care database. npj Primary Care Respiratory Medicine. 2017;**27**:58

[11] Bachasson D, Wuyam B, Pepin J-L, Tamisier R, Levy P. Quadriceps and respiratory muscle fatigue following high-intensity cycling in COPD patients. PLoS ONE. 2013;**8**(12):e83432

[12] Schols AM, Ferreira IM, Franssen FM. Nutritional assessment and therapy in COPD: A European Respiratory Society statement. The European Respiratory Journal. 2014;**44**:1504-1520

[13] Xiong Y, Wang M, Xiao H. Echo intensity of the rectus femoris in stable COPD patients. International Journal of COPD. 2017;**12**:3007-3015

[14] Benedika B, Farkasb J. Mini nutritional assessment, body composition, and hospitalizations in patients with chronic obstructive pulmonary disease. Respiratory Medicine. 2011;**105**(S1):S38-S43

[15] Kornum M, Nørgard C. Obesity and risk of subsequent hospitalization with pneumonia. The European Respiratory Journal. 2010;**36**:1330-1336

[16] Marcellis E. Exercise capacity, muscle strength and fatigue in sarcoidosis. The European Respiratory Journal. 2011;**38**:628-634

[17] Bonsignore WT, Nicholas JM, Montserrat M, Eckel J. Adipose tissue in obesity and obstructive sleep apnea. The European Respiratory Journal. 2012;**39**: 746-767

[18] Marmorato DM. Study of peripheral muscle strength and severity indexes in individuals with chronic obstructive pulmonary disease. Physiotherapy Research International: The Journal for Researchers and Clinicians in Physical Therapy. 2010:223-242

[19] Abe T, Counts BR, Barnett BE. Associations between handgrip strength and ultrasound-measured muscle thickness of the hand and forearm in young men and women. Ultrasound in Medicine & Biology. 2015;**41**:2125-2130

[20] Devid B, Brønstad E. High-intensity knee extensor training restores skeletal muscle function in COPD patients. The European Respiratory Journal. 2012;**40**: 1130-1136

[21] Vogiatzis S. Strategies of muscle training in very severe COPD patients. The European Respiratory Journal. 2011; **38**:971-975

[22] Shrikrishna D, Patel M. Quadriceps wasting and physical inactivity in patients with COPD. The European Respiratory Journal. 2012;**40**:1115-1122

[23] Maddocks M, Shrikrishna D, Vitoriano S, Natanek SA, Tanner RJ, Hart N. Skeletal muscle adiposity is associated with physical activity, exercise capacity and fiber shift in COPD. The European Respiratory Journal. 2014;**44**:1188-1198

[24] Spencer LM, Alison JA, Keough ZJM. Maintaining benefits following pulmonary rehabilitation:

A randomized controlled trial. The European Respiratory Journal. 2010;**35**: 571-577

[25] Soicher JE, Mayo NE. Trajectories of endurance activity following pulmonary rehabilitation in COPD patients. The European Respiratory Journal. 2012;**39**: 272-278

[26] Marciniuk DD, Brooks D. Optimizing pulmonary rehabilitation in chronic obstructive pulmonary disease-practical issues: A Canadian Thoracic Society Clinical Practice Guideline. Canadian Respiratory Journal. 2010; **17**(4):159-168

[27] Hill K, Vogiatzis I, Burtin C. The importance of components of pulmonary rehabilitation, other than exercise training, in COPD. European Respiratory Review. 2013;**22**:405-413

[28] Licker M, Schnyder JM. Impact of aerobic exercise capacity and procedure-related factors in lung cancer surgery. The European Respiratory Journal. 2011;**37**:1189-1198

[29] Ramon MA, Gimeno-Santos E, Ferrer J, Balcells E, Rodrıguez E, Batlle J, et al. Hospital admissions and exercise capacity decline in patients with COPD. European Respiratory Journal. 2014;**43**: 1018-1027

[30] Savarese G, Musella F, D'Amore C, Losco T, Marciano C, Gargiulo P, et al. Hemodynamics, exercise capacity and clinical events in pulmonary arterial hypertension. The European Respiratory Journal. 2013;**42**:414-424

[31] Jones SE, Maddocks M, Kon SS. Sarcopenia in COPD: Prevalence, clinical correlates and response to pulmonary rehabilitation. Thorax. 2015; **70**:213-218

[32] Spruit MA, Singh SJ, Garvey C. An official American Thoracic Society/

European Respiratory Society
statement: Key concepts and advances
in pulmonary rehabilitation. American
Journal of Respiratory and Critical Care
Medicine. 2013;**188**:222-230

[33] Jackson AS, Shrikrishna D. Vitamin
D and skeletal muscle strength and
endurance in COPD. The European
Respiratory Journal. 2013;**41**:309-316

[34] Puhan MA, Siebeling L, Zoller M,
Muggensturm P, Riet G. Simple
functional performance tests and
mortality in COPD. The European
Respiratory Journal. 2013;**42**:956-963

[35] Mainguy V, Malenfant S,
Neyron AS. Repeat ability and
responsiveness of exercise tests in
pulmonary arterial hypertension. The
European Respiratory Journal. 2013;**42**:
425-434

[36] Gloeck R, Marinov B, Pitta F.
Practical recommendations for exercise
training in patients with COPD.
European Respiratory Review. 2013;**22**:
178-186

[37] Celli CBR, Barria P, Casas A,
Cotee C, Torres JP. The 6-min walk
distance in healthy subjects: Reference
standards from seven countries. The
European Respiratory Journal. 2011;**37**:
150-156

Section 3

Respiratory System and Environment

Chapter 4

Environmental Media and Associated Respiratory Defects

Ibiwumi Saliu and Evangelisca Akiomon

Abstract

Environmental media majorly connotes abiotic components of the natural environment, namely, air, water and soil. Pollution to these media has resulted to a great deal of respiratory defects. Epidemiologic studies conducted in the U.S. and abroad provide evidence of associations between short and long-term exposure to fine particles in the air and both decrements in lung function growth in children and increased respiratory symptoms. Particles deposited in the respiratory tract in sufficient amounts induce lung inflammation, which has been demonstrated in both animal and controlled human exposure studies. More recently, the International Agency for Research on Cancer (IARC) also conducted an evaluation on the carcinogenicity of outdoor air pollution in the respiratory tract, including particle pollution, and concluded that both are Group I agents (carcinogenic to humans). Air pollution has been given great priority as a causal factor for respiratory defects; meanwhile dust particles from contaminated soil could also cause a great havoc. Moreover polluted water is also a major causal pathway. According to world health organization (WHO) 80% diseases are waterborne. Though water is an important natural resource used for drinking and other developmental purposes in our lives but health risk associated with polluted water includes different diseases in which respiratory diseases are the major ones. Bacterial, viral and parasitic diseases are spreading through polluted water and affecting human health. Poliomyelitis virus is responsible for poliomyelitis, sore throat, fever, nausea, which are all due to polluted water.

Keywords: air, water, soil, lung inflammation, cancer, nausea

1. Introduction

Defect is any abnormality or imperfection that is capable of impairing quality, function or utility [1]. A respiratory defect is therefore any impairment in the vital function or utility of the respiratory system caused by abnormalities (congenital or induced) in one or more of the respiratory organs. The major organs of the respiratory system are the nose and nasal cavity, larynx, pharynx, bronchi, alveoli, trachea and the lungs. They function primarily to provide oxygen to body tissues for cellular respiration, remove the waste product carbon dioxide, and help to maintain acid-base balance [2]. Due to the anatomy of the respiratory tract, it is constantly exposed to microbes, which is why the respiratory system includes many mechanisms to defend itself and prevent pathogens from entering the body.

Lungs are the powerhouse of the respiratory system. Through inhalation oxygen is brought into the body with and exhalation rids the body of carbon dioxide. Harm to the lungs can result from presentation to numerous things noticeable all around, for example, allergens, lethal substances, metals, and molds. Research has indicated that long term introduction to air contaminations can influence the development and improvement of the lungs, and increment the danger of creating asthma, emphysema, and other lung illnesses.

The NIEHS-sponsored Harvard Six Cities Study found a significant relationship between introduction to fine particle air pollution and early deaths. Likewise, perhaps the greatest risk to lung wellbeing is tobacco smoke, containing many poisons, for example, nicotine carbon monooxide, tar, arsenic, cadmium, methane, etc. [3]. Respiratory disease is a common and significant cause of illness and death around the world. The most common cause of illness in children in developed countries and a leading cause of death in children in developing areas are diseases of the lung and airways. According to the Centers for Disease Control and Prevention (CDC), chronic obstructive pulmonary disorder (COPD) is the fourth leading cause of death in the United States. Its prevalence increases with age. Men are more likely to have the disease, but the death rate for men and women is about the same. In another research conducted in the United States in 2010, it was discovered about 6.8 million visits to the emergency division were for respiratory issues from patients younger than 18. About one-seventh of the United Kingdom population is affected by some type of persistent lung disease, most commonly COPD, which incorporates asthma, persistent bronchitis, and emphysema. There are various ways by which respiratory disorders can be classified: it could be based on

1. The organ or tissue involved

2. Pattern of signs and symptoms associated with the disease

3. Causal factor of the disease

Generally, respiratory disorders can be classified into these areas;

- Obstructive respiratory conditions are usually characterized by obstructed airflow inflamed and easily collapsible airways. Examples are asthma, bronchitis, emphysema, bronchiectasis, etc.

- Restrictive respiratory conditions these disease restrict lung expansion resulting in reduced lung volumes, either because of an alteration in lung parenchyma or because of a disease of the pleura (e.g., fibrosis, sarcoidosis, alveolar damage, pleural effusion).

- Vascular diseases. They are diseases that affect the blood vessels from the heart to the lungs. They are of two major types: pulmonary embolism and pulmonary hypertension.

- Environmental and occupational respiratory "diseases." These respiratory defects are caused by harmful particles, mists, vapors, or gases that are inhaled, in the surroundings where people live and usually while people work. Various sorts of particles elicit various reactions in the body. For instance, particles like animal dander could cause unfavorably susceptible responses, like hay fever-like symptoms or a kind of asthma. Different particles cause hurt not by activating hypersensitive responses, however by being harmful to the cells of the aviation routes and air sacs in the lung. A few particles, for example, quartz residue and asbestos, may cause incessant aggravation that can prompt scarring of lung tissue (aspiratory fibrosis). Certain lethal particles, for example, asbestos, can cause lung malignant growth, especially in people who smoke, or cancer of the lining of the chest and lung (mesothelioma), (e.g., pneumonia, tuberculosis, asbestosis, particulate pollutants) [4, 5].

2. Overview of the environment and its various components (media)

The environment can be defined as all that which is external to the individual host. It can be divided into physical, biological, social, and cultural factors, any or all of which can influence health status in populations [6]. In other words environment refers to those things that surrounds living beings and affect their lives. The environment is man's basic life support system. It provides the air we breathe, the water we drink, the food we eat and the land where we live.

The four major components of environment include lithosphere, hydrosphere, atmosphere and biosphere, corresponding to rocks, water, air and life respectively.

Lithosphere is the outermost layer of earth called crust, consisting of various minerals, it is where we have the rocks and soil. It is about 100 km in depth, and is found on both land and seas. The principal component of lithosphere is earth's tectonic plates.

Hydrosphere, covering 70% of earth's surface, involves all the varieties of water bodies on earth such as seas, oceans, rivers, lakes, lakes, streams, etc. 97.5% of water found on earth is in the seas as salt water. Just 2.5% of water on earth is freshwater. Out of this, 30.8% makes up the rivers, repositories, and lakes and is easily accessible to man.

Atmosphere is the gaseous, envelope-like material encasing the earth. It is unique to the earth because its abundance of oxygen (20.95%) which is vital in supporting life on earth. other constituents of the atmosphere includes 78.08% nitrogen, 0.93% argon, 0.038% carbon dioxide, and trace elements like, hydrogen, helium, noble gases and water vapor.

Biosphere refers to all the regions on Earth where life exists. The ecosystems that support life could be in soil, air, water, or land.

3. The environment and health

The environment we live in is a major determinant of our state of health, from the rural zones to thick urban communities, the sort meals we eat and water we drink, places we live to the spots we work, and hence harm to our indigenous habitat, likewise brings about harm to human wellbeing. Factors, like inaccessible, safe drinking water and sanitation, air pollution and changes in climatic conditions contribute to 23% of all deaths globally and 36% of all deaths among youngsters between the ages of 0 and 14. It is estimated that 1.8 billion individuals get drinking water from fecally contaminated water sources and 2.5 billion individuals live without essential sanitation facilities, increasing the rate of diarrheal infections, malnutrition and deaths. Insufficient water for hygienic purposes, consumption of unsafe water and absence of access to sanitation together makes up about 88% of deaths from diarrheal diseases, resulting in 11% of mortality of kids younger than 5 [7].

Health impacts associated with damage to our environment are numerous and diverse ranging from diarrheal and vector-borne diseases to respiratory diseases, ischemic heart disease and stroke to mental health impacts of extreme weather events, failing livelihoods, conflict and displacement [8]. This therefore implies that in order to reduce the incidence of disease more attention needs to be paid to reducing the environmental causes such as:

Poor air quality: about 3 billion people depend on wood, charcoal, animal compost, and crop waste as fuel for the supply of energy in households. Fifty to seventy percent of Africa's populace cooks with solid fuel, majorly exposing women and children to massive quantities of pollutants, in various concentrations, posing considerable health risks to humans, hence increasing the risk of diseases like pneumonia and chronic respiratory disorders on such exposures.

Poor sanitation diseases such as diarrhea result from poor sanitary conditions and unsafe water and are responsible for the deaths of 1.8 million people every year amounting to an estimated 4.1% of the total DALY global burden of disease. 88% of this disease burden is attributable to unsafe water supply, poor sanitation and poor hygiene practices. Provision of safe drinking water and sanitation are important measures of improvement.

Poor landscape and urban land management poor landscaping and lack of urban and regional planning contribute to overpopulation and overcrowding. This will increase the spread of infectious diseases, favor the proliferation of pest, poses a lot of pressure on the basic amenities and infrastructure, increased areas of stagnant water and thereby increasing the risk of malaria. Across the globe, about 30 million cases of malaria are being recorded yearly, leading to over a million deaths, with approximately 90% of these deaths occurring majorly among young children in Africa. Malaria makes up 10% of Africa's overall disease burden and is the main cause of death among children under the age of five. Africa bears over us$ 12 billion in lost GDP yearly due to malaria.

Overexploitation and degradation of natural resources: a major factor contributing to food insecurity is over exploitation and degradation of natural resources, it reduces the capacity of the land to produce crops and sustain livestock. The resultant effect of food shortage is malnutrition and in worst cases, starvation. Asides reducing immunity to other diseases, severe cases of malnutrition also causes stunting in children and hinders healthy development. According to World Health Organization, across the globe, one in seven people are affected by hunger. Also, of the 10.9 million child deaths each year, half is attributable to malnutrition+ [7].

4. Air-induced respiratory disorders

Air is a mixture of gases comprising the atmosphere, living things requires air for their existence. Thus the quality of air we breathe in is a vital indicator of our state of health. When we breathe in polluted air, the pollutants get inhaled deep into the lungs, resulting in serious harm to the respiratory tract. Polluted air can set off new cases of asthma, worsen existing cases of respiratory diseases and trigger the development or advancement of chronic illness including lung cancer, chronic obstructive pulmonary disease, and emphysema. Air pollutants also have a considerably adverse effect on lung development, thus, increasing the vulnerability for developing lung diseases in the long run [9]. Burning of wood, animal waste, and crops (biomass fuel) is an important source of particulate matter indoors in developing countries, Secondhand smoke is also an important source of indoor air pollution [10]. Air pollution-related lung disease increases the risk of heart and blood vessel disorders and may increase the risk of lung cancer. Long-term exposure to air pollution may increase respiratory infections and symptoms of respiratory disorders (such as cough and difficulty breathing) and decrease lung function.

Ozone, a primary air pollutant and basic constituent of smog, is a strong lung irritant. During the year, levels are noticeably higher during summer likewise in contrast with other period of the day, ozone levels tend to be higher, late in the morning and early in the afternoon. Short-term exposures can cause breathing difficulties, chest pain, and airway hyperactivity. Participating in outdoor activities on days when ozone pollution is high can increase the risk of developing asthma among children. Long-term exposure to ozone causes a small, permanent decrease in lung function. When fossil fuel that are high in sulfur content are burnt, they acidic particles called sulfur oxides, which are deposited in the upper respiratory tract and can cause constriction and irritation resulting in increased risk of chronic bronchitis and symptoms like breathing difficulty. Particles affect the lung in various ways, depending on their primary material. Particles of the same material also may have different effects depending on their size and shape. Variation in environmental conditions and location affects the level of pollutants in the air. For instance, on warm and humid days, ozone tends to remain in the air. Carbon monoxide levels also tend to be high during peak periods when many commuters drive to or from work. The Air Quality Index is utilized to convey how polluted the air is at a given point in time. Individuals, particularly those with heart or lung disorders, can utilize the Air Quality Index to manage their choice of outdoor activities on days when pollution levels are high [10].

AQI values	Air quality condition
0–50	Good
51–100	Moderate
101–150	Unhealthy for members of sensitive groups
151–200	Unhealthy
201–300	Very unhealthy
301–500	Hazardous

Source: msdmanuals.com.
Adapted from US Environmental Protection Agency: Air Quality Index: A guide to air quality and your health. Research Triangle Park, NC, 2009.

Air Quality Index.

Children are more vulnerable to the impacts of air pollution. They inhale through their mouths, bypassing the sifting impacts of the nasal entries and permitting pollutants to travel further into the lungs. They have a larger lung surface area comparative with their weight and breathe in generally more air, compared to grown-ups. Children also spend more time outdoors especially in the afternoon and throughout the summer months when ozone and other pollutant levels are at their peak and may disregard early indications of impact of air pollution such as, an asthma exacerbation leading to severe attacks. A combination of these with the adverse impact of some pollutants on lung development and the immaturity of children's enzyme and immune systems that detoxify pollutants, results in series of factors that contribute to children's increased sensitivity to air pollutants [9].

Asthma is a chronic disease which affects the bronchi and bronchioles of the lungs characterized by inflammation and narrowing of the airways, it causes a sensation of tightness in the chest, shortness of breath, wheezing, and coughing. Over 20 million people in the U.S., including six million children now gasp for breath due to asthma [9]. Asthma triggers are numerous and vary; dust, smoke, pollen, and volatile organic compounds are examples of asthma triggers. Primary pollutants such as ozone, carbon monoxide, sulfur dioxide, and nitrogen oxides are also outdoor triggers.

Chronic Obstructive Pulmonary Disease (COPD), chronic bronchitis and emphysema. Chronic Obstructive Pulmonary Disease (COPD) is an irreversible condition caused by exposure to pollutants that produce inflammation, and immunological response. It is characterized by narrowing of the airways. In larger airways, the inflammatory response is referred to as chronic bronchitis. It leads to tissue damage or emphysema in the tiny air cells at the end of the lung's smallest passageways [9]. Emphysema is a chronic disease that causes reduction of the respiratory surface due to the damage to the lung alveolar walls. It is caused mainly by cigarette smoking, dust, chemicals and exposure to passive cigarette smoking. The main symptoms of emphysema include shortness of breath and cough. Emphysema might lead to a loss of elasticity of the lungs.

Lung cancer in the U.S, the main cancer killer in both men and women is lung cancer, is frequently (and precisely) related with smoking tobacco. While that is valid, there are also other risk factors associated with lung cancer, along with air pollution. Particulate issue and ozone are also implicated in mortality due to lung cancer [9].

5. Occupational respiratory disorders

Occupational respiratory disorders are defined as any disorder which affects the respiratory system by long-term inhalation of chemicals, proteins, and dust. Occupational respiratory disorders might happen due to the inhalation of the following substances fumes from metals, smoke from burning organic materials, sprays of varnishes, paint, acids, and pesticides, dust from cotton, silica, coal, drug powders and pesticides and gases from industries. The type of occupational respiratory disorder depends on the environment to which the person is exposed: people, particularly those with other lung disorders, are at risk when they are exposed to air pollution in the environment or to contaminants in indoor environments. Some many more people are at risk of occupational asthma as a result of exposure in the workplace. Exposure to asbestos can cause asbestosis, mesothelioma, and asbestos-related pleural disease. People who work with beryllium, such as aerospace workers, are at risk of beryllium disease. Byssinosis is prevalent among people who work with cotton, flax, or hemp. Coal workers and graphite workers are at risk of coal workers' pneumoconiosis. Prolonged exposure to silica would result in silicosis [5].

6. Soil-related respiratory defects

Soil is a complex system of air, water, minerals, natural matter and biota that covers the terrestrial earth in layers over the underlying bedrock. Endogenous segments of soil incorporate minerogenic colloidal clay and trace elements, biogenic organic materials, and biota, any of which might be advantageous, impeding or harmful, contingent upon their relative concentrations and the exposure pathway. Soils are also significant source of supplements, and they go about as common channels to expel contaminants from water [11]. Notwithstanding, soils may contain heavy metals, chemicals, or pathogens that can adversely affect human wellbeing. As soil is persistently being made airborne and afterward dispersed through the air by the global components of climate and weather. Soil, in dust forms, can sometime be dispersed to great distances therefore, increasing the exposure levels of humans to soil particles throughout their evolutionary history, from both local and regional sources and potentially from almost anywhere on the planet [12].

The NIH National Institute of Allergy and Infectious Diseases also reported that airborne residue is the main source of environmental agents that encourage human allergic disorders [13]. The basis of reactions of the human immune system to airborne dusts has created an outcome of long term exposures, on a developmental time scale to minerogenic, biogenic and anthropogenic parts that are pervasive segments of the natural history of people. The manner in which the respiratory system reacts to inhaled particles depends on where the molecule settles. For instance, irritant dust that settles in the nose may prompt rhinitis, an irritation of the mucous layer. Peradventure the molecules attacks the larger airways, irritation of the trachea (tracheitis) or the bronchi (bronchitis) might be seen (**Figure 1**).

The most prominent responses of the lung happen in the core of the organ. Particles that avoid being eliminated in the nose or throat in turn settles in the sacs or near the distal part of the airways. Be that as it may, if the dust quantity is enormous, the macrophage system can fail. Dust particles and residue containing macrophages accumulate in the lung tissues, causing damage to the lungs. The amount of dust and the types of particles involved is proportional to how severe the lung damage will be. For instance, after the macrophages swallow silica particles, they undergo apoptosis and emit harmful substances. These substances cause sinewy or scar tissue to form. This tissue is simply the body's typical means of fixing itself. However, on account of crystalline silica so much stringy tissue and scarring structures accumulate so much that the lung capacity can be debilitated. The general name for this condition for sinewy tissue arrangement and scarring is fibrosis. The particles which cause fibrosis or scarring are called fibrogenic. At the point when fibrosis is brought about by crystalline silica, the condition is called silicosis. Once deposited in the pulmonary alveoli, these quartz-based minerogenic silica particles start a fibrotic wound reaction that can in the long run lead to silicosis, a debilitating pneumonic condition. Extreme silicosis can additionally harm the immune system by impeding the ability of macrophages in the lungs to distort the growth of pathogenic organisms found in airborne dust, prompting a variety of bacterial infections [14].

An inherited hypersensitivity of the immune system to react to a specific ubiquitous airborne antigen or the presentation to a new aerosoled soil material

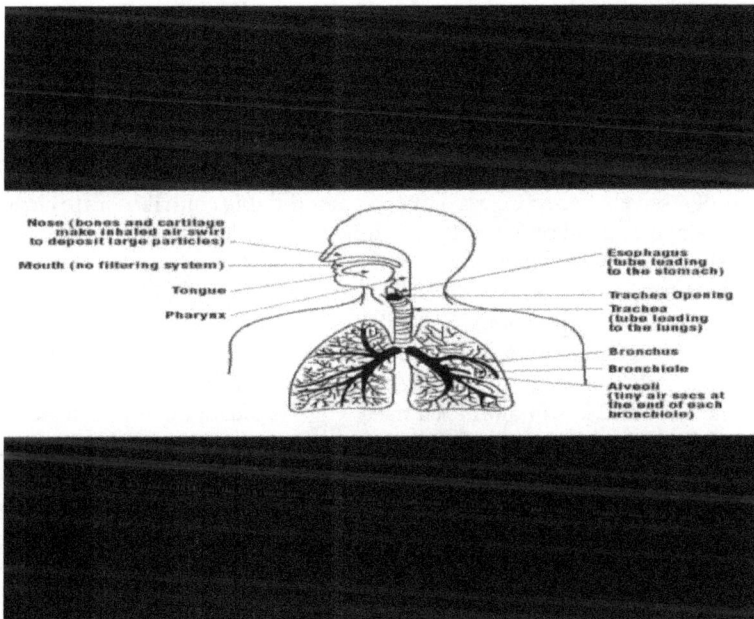

Figure 1.
The human respiratory tract (source: Pinterest.com).

may trigger reactions of the immune system that can bring about asthma and other related conditions. The scarring effect of inhaling minerogenic dust upon the nasopharyngeal mucosa creates an ideal environment for contamination of *N. meningitis*. Studies have shown that samples of aerosoled materials collected during dust events on the African continent have a host of human pathogens, including *Actinobacter calcoaceticus*, *Kocuria rosea* and others [15].

7. Water-related respiratory defects

Varieties of water sources can become repositories and vectors for contaminants related with intense and chronic lung infection. Improper sewage sanitation is an issue for about 40% of the world's population, and a millions of individuals die every year from waterborne illnesses [16]. Polluted and filthy water is exceptionally harmful for living organism particularly for wellbeing of humans. It causes numerous health issues which can eventually lead to death if not treated on time. Inhalation of contaminated aerosols is the most important route of exposure leading to water-related lung disease. Nonetheless, dermal assimilation, dispersed marine-acquired wound infections, and ingestion or aspiration of water containing harmful contaminants have been related with pneumonic sickness also. Upper respiratory side effects are common after water exercises, with over 40% of recreational swimmers reporting sinus symptoms. Hyperemia is the suspected etiology, and indication of prolonged water exposure [17].

The center for disease control (CDC) likewise detailed the peril in swimming; swimmers are said to be in danger of respiratory diseases if they take in steam or mist from a pool or hot tub that contains harmful germs. A respiratory sickness brought about by the germ Legionella is one of the most reoccurring waterborne ailments (drinking water and recreational water). If legionnaires' disease develops and is unrecognized, mortality might be as high as 10%. Untreated lakes and streams are said to have been the culprit, just as public pool or hot tub [14].

8. Food-related respiratory defects

Diet and nutrition might be significant modifiable risk factors for the improvement, movement and the management of obstructive lung infections, for example, asthma and incessant obstructive pneumonic ailment (COPD) [18]. Diet and nutrition are progressively becoming recognized as modifiable contributors to chronic disease development and progression. Significant proof has shown the essence of dietary intake in obstructive lung disease, for example, asthma and incessant obstructive pneumonic disease (COPD) in both early life and ailment advancement. This dietary example should comprise of a high admission of insignificantly handled plant nourishments, to be specific; organic product, vegetables, breads, grains, beans, nuts and seeds, low to direct intake of dairy nourishments, fish, poultry and wine and low intake of red meat. High intake of olive oil brings about a dietary synthesis that is low in soaked fat however still moderate in complete fat. Be that as it may, over sustenance and coming about weight are unmistakably connected with asthma; however the instruments included are still under scrutiny [19].

9. Recommendation and conclusion

The ability to survey the environment and exposure must be improved upon so as to comprehend the impact environmental factors have on disease and to decide if new ecological variables may bring about illness. Assessment methods are required to screen an individual's absolute exposure to environmental factors over a lifetime as opposed to during a specific time span or in one circumstance. This evaluation could be accomplished at any rate halfway with the improvement of biomarkers that indicate exposure to precipitating factors from in utero to the end of life. Research surveys that address the multifaceted nature of the exposures are very likely to show the impact of environmental factors on lung infection [20].

So likewise, more discoveries should be made about the interaction between the individual and environment to have a better definition of at risk population. These efforts ought to not only distinguish most at risk population but to likewise consider how alterations of environmental factors could decrease the impact of disease. An integrative methodology will be required for these researches, alongside dependence on genetic technologies, bioinformatics, and complex biostatistical techniques. Likewise distinguishing hereditary components related with risk of exposure. This research could also recognize biomarkers of illness and characterize potential pathogenic pathways that might be focused to lessen or treat sickness [20].

Furthermore, the means by which environmental toxins influence disease development should be characterized. In spite of the fact that it is confirmed that indoor air pollution increases the danger of cardiovascular illness, so also indoor air contamination due to biomass smoke accumulation especially in children, the mechanisms by which these toxicants act are unknown [20].

Finally, to mitigate environmental lung disease, multidisciplinary research and public health programs are expected to decipher what is found about these toxicants and pathways for modifications in the environment in order to help individuals who are in danger of respiratory diseases. Presently, there are not too many researchers and clinicians who have the interest and capacity to lead environmental research. Hence, a significant step to improve in this field is to prepare more analysts. With these methodologies and the improvement of organizations among researchers and the general population everywhere, the role of environmental factors in lung disease will keep on being characterized and techniques to forestall must be implemented [20].

To avoid respiratory problems with other health issues especially in work areas caused by exposure to dust, controls must be implemented. According to the hierarchy of control, the main consideration should be hazardous substances substituted with non-hazardous substances. Where substitution is impossible, other engineering control methods could be introduced. which include: the use of wet processes for the enclosure of dust-producing processes under negative air pressure (slight vacuum compared to the air pressure outside the enclosure); exhausting air containing dust through a collection system before releasing to the atmosphere; use of vacuums instead of brooms good housekeeping efficient storage and transport controlled disposal of dangerous waste; use of personal protective equipment is also very important, but it should nevertheless be the last resort of protection. Personal protective equipment should not be a substitute for proper dust control and should be used only where dust control methods are not yet effective or are inadequate. Most importantly, workers, through education and proper orientation, must comprehend the need to avoid the risks of dust [18].

Author details

Ibiwumi Saliu[1,2]* and Evangelisca Akiomon[1]

1 Department of Environmental Health Sciences, Faculty of Public Health, University of Ibadan, Nigeria

2 Blue-Gate Public Health Initiative, Ibadan, Oyo State, Nigeria

*Address all correspondence to: saliuibiwumi@yahoo.com

IntechOpen

References

[1] Available from: https://www.merriam-webster.com/dictionary/defect

[2] The Biology of Aging. Available from: https://courses.lumenlearning.com/atd-herkimer-biologyofaging/chapter/organs-and-structures-of-the-respiratory-system/

[3] Douglas WD. Health effects of particulate air. Annals of Epidemiology. 2009;**19**(4):257-263. DOI: 10.1016/j.annepidem.2009.01.018. Available from: http://www.ncbi.nlm.nih.gov/pubmed/8179653

[4] Available from: https://www.disabled-world.com/health/respiratory/

[5] Lara AR. Overview of Environmental Lung Diseases. Available from: https://www.msdmanuals.com/home/lung-and-airway-disorders/environmental-lung-diseases/overview-of-environmental-lung-diseases

[6] Last JM, editor. A Dictionary of Epidemiology. 3rd ed. New York: Oxford University Press; 1995

[7] Centre for Disease Control and Prevention. Global Water, sanitation, & Hygiene (WASH) Fast Facts. 2013. Available from: http://www.cdc.gov/healthywater/global/wash_statistics.html#one

[8] Environment and Health. Available from: https://www.irishaid.ie/media/irishaid/allwebsitemedia/20newsandpublications/publicationpdfsenglish/environment-keysheet-8-health.pdf

[9] How Air Pollution Contributes to Lung Diseases. Physicians for Social Responsibility. Available from: https://www.psr.org/wp-content/uploads/2018/05/air-pollution-effects-respiratory.pdf

[10] Lara AR. Air Pollution Related Illness. Available from: https://www.msdmanuals.com/home/lung-and-airway-disorders/environmental-lung-diseases/air-pollution%E2%80%93related-illness

[11] United States Environmental Protection Agency. Learn the Issues. Available from: http://www.epa.gov/epahome/learn.htm

[12] World Health Organization. The World Health Report 2002—Reducing Risks, Promoting Healthy Life. Available from: http://www.who.int/whr/2002/en

[13] National Institute of Allergy and Infections Diseases. Available from: http://www.niaid.nih.gov

[14] Centers for Disease Control and Prevention. Surgeon General's Report—The Health Consequences of Involuntary Exposure to Tobacco Smoke. 2006. Available from: http://www.cdc.gov/tobacco/data_statistics/sgr/2006/index.htm

[15] National Institutes of Environmental Health Sciences—National Institutes of Health. Environmental Health Topics: Conditions & Diseases. Available from: http://www.niehs.nih.gov/health/topics/index.c

[16] Massoud MA, Tarhini A, Nasr JA. Decentralized approaches to wastewater treatment and management: Applicability in developing countries. Journal of Environmental Management. 2009;**90**:652-659. DOI: 10.1016/j.jenvman.2008.07.001

[17] Fazel-ur Rehman M. Journal of Medicinal and Chemical Sciences Review Article Polluted Water Borne Diseases: Symptoms, Causes, Treatment and Prevention

[18] Canadian Centre for Occupational Health & Safety. 2019

[19] Caldefie-Chezet F, Poulin A, Vasson MP. Leptin regulates functional capacities of polymorphonuclear neutrophils. Free Radical Research. 2003;**37**:809-814

[20] Crapo JD, Broaddus VC, Brody AR, Malindzak G, Samet J, Wright JR, et al. Workshop on lung disease and the environment: Where do we go from here? American Journal of Respiratory and Critical Care Medicine. 2003;**168**:250-254

Chapter 5

Historical Aspects of Hyperbaric Physiology and Medicine

Chandrasekhar Krishnamurti

Abstract

The history of hyperbaric oxygen therapy (HBOT) makes for fascinating reading. From pneumatic chambers and compressed air baths to empirical therapeutic applications during the nineteenth century, the impetus to scientific application of HBOT began in seeking solution for decompression sickness during various construction ventures. French physiologist Paul Bert's research was pathbreaking and provided a scientific explanation on the etiology of the "bends." In 1908, JS Haldane's experiments recommended staged decompression and made diving safe. In 1921, OJ Cunningham employed HBOT to treat hypoxia secondary to lung infections successfully. It was cardiac surgeon Ite Boerema who put HBOT on a solid footing with his open-heart surgery results in various pediatric cardiac conditions and rightly deserved the title of father of modern-day hyperbaric medicine. From 1937 onwards, HBOT research snowballed into treating a wide variety of diseases. In 1999, the Undersea and Hyperbaric Medical Society and Food and Drug Administration recognized the value of HBOT, and this led to its becoming a major tool in the armamentarium of clinicians, either as a primary or adjunctive therapy for a spectrum of diseases.

Keywords: history, hyperbaric oxygen therapy

1. Introduction

Since 4500 BC, breath-holding dives for mother-of-pearl, sea sponges, and coral was a distinct occupation. These free divers could hold their breath for extended periods of time, and their work was confined to waters less than 30 m (100 ft) deep. It was undoubtedly a hazardous occupation, and many of them succumbed to decompression sickness after rapid surfacing. Persian king Xerxes the Great (520–465 BC) employed divers to salvage sunken goods and treasures from the wrecks of Greek ships he had sunk in numerous battles at sea. Some of these dives were recorded to depths of 20–30 m and lasting 4 minutes at a time. The ancient Greek historians Herodotos and Pausanias wrote about a Greek hero named Scyllias from Scione, who used a reed and diving capsule made from animal skins to cut the mooring lines of enemy ships. Pausanias even taught his own daughter Hydna to dive. Alexander the Great (365–323 BC), under the advice of a reputed astronomer named Ethicus, dived into the Bosphorus straits in a bathysphere, accompanied by a dog, a cat, and a rooster, after entrusting the security of the hoisting chain to his most loyal mistress. Taking advantage of the moment, she chose to elope with her lover after casting the chain into the sea, abandoning Alexander and leaving him to figure out his escape on his own! In 300 BC, Aristotle described the ruptured eardrum as a complication of undersea diving.

Figure 1.
Henshaw and his domicilium.

While living in Venice in the late fifteenth century, Leonardo da Vinci designed diving suits to enable divers cut holes in the hulls of invading ships, but none seem to have been developed or used [1].

In 1620, Dutch inventor Cornelis Jacobszoon Drebbel (1572–1633) designed a wooden diving boat, sealed against water by greased leather, to travel in the River Thames at a depth of around 4 m, from Westminster to Greenwich. Air was supplied by two tubes with floats to maintain one end above water [2]. In sixteenth century England and France, full diving suits made of leather were used to depths of 60 ft with air being pumped down from the surface with the aid of manual pumps.

The first documented use of hyperbaric therapy was in 1662, when a British clergyman and physician named Nathaniel Henshaw used a system of organ bellows with unidirectional valves to change the atmospheric pressure in a sealed chamber called a domicilium (**Figure 1**). Without any scientific rationale whatsoever, Henshaw claimed that high air pressures would remedy acute conditions while lower pressures would yield salutary results in chronic disorders. His domicilium therapy was touted to improve digestion and prevent lung diseases by manipulating ambient pressures without increasing oxygen concentrations, as oxygen was not discovered until nearly a century later [2].

In 1690, Edmond Halley designed a diving bell ventilated with weighted barrels of air sent down from the surface. Employing this device, Halley, escorted by five of his close friends, undertook a dive to a depth of 60 ft in the River Thames in that bell and remained submerged at that depth for 90 minutes. Too heavy for salvage work, Halley made improvements to his bell, extending his underwater exposure time to over 4 hours. The first deep-sea diving suit was invented in 1819 by Augustus Siebe. It used compressed air supplied to the helmet for ease of movement underwater.

All of these early submersibles used ambient air and were called "pneumatic chambers" or "compressed air baths."

2. The era of empirical HBOT/HBO spas

Nearly two centuries later, in the 1830s, there was a rebirth of interest in hyperbaric medicine in France. In 1834, the French physician Junod built a hyperbaric chamber designed by the steam engine inventor James Watt, who was well-versed in pressure physics. This appliance could generate a maximum of 4 atmospheres pressure and used to treat pulmonary afflictions using pressures

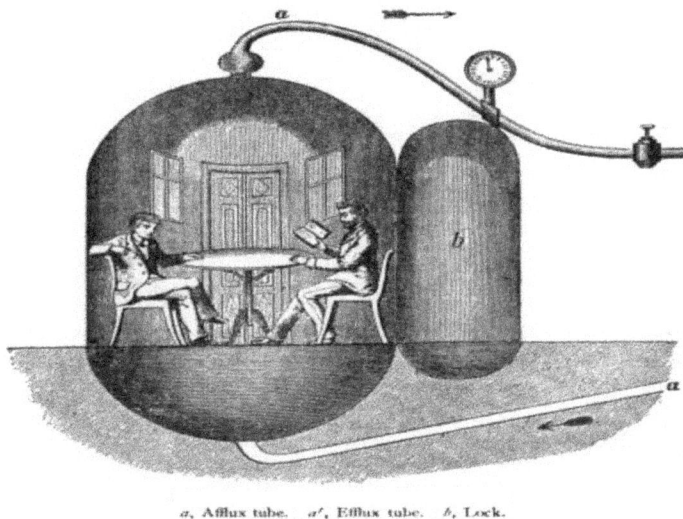

a, Afflux tube. *a'*, Efflux tube. *b*, Lock.

Figure 2.
Taberie's pneumatic chamber.

between 2 and 4 ATA. Junod referred to his treatment as "Le Bain d'air comprimé" (the compressed air bath), and claimed that it increased circulation to the internal organs and the brain, resulting in feelings of well-being and better general health.

Taberie designed a spherical pneumatic chamber made of cast iron with two pipes, one to provide pressure from a hydraulic compressor run by steam and the other to allow for ventilation. Carpet covered the floor to conceal the pipes, and it featured an antechamber to allow the physician to enter and exit without disturbing the pressure. The passage was also used to stock books, newspapers, and drinks for the patients (**Figure 2**).

Lange had a cylindrical chamber constructed out of wrought iron, designed to accommodate four persons. The temperature of the compressed air within the chamber was lowered in two ways. The first employed a stream of cold water directed against the force pump and the supply pipes. The second method was by filling a cup-shaped space at the top of the chamber with cold water and allowing it to cascade down the sides to soak sheets of linen and cool the air by evaporation. In winter the chamber was kept at a comfortable temperature by heating. The chamber was also provided with a device for regulating the flow of the incoming air so that it entered in a steady stream (instead of a succession of puffs in earlier versions) by a force pump. The pressure was secured, as in Tabarie's system, by regulating the inflow and outflow of the air (**Figure 3**).

Leibig's pneumatic chamber was located at Dianabad in Reichenhall, Bavaria, Germany. This pneumatic chamber had three chambers, each one capable of accommodating three persons. One antechamber connected all three rooms, allowing the physician to enter and exit without affecting the ambient pressure. The antechamber also acted as a large pressure regulator, preventing the patients from being affected by sudden surges of pressure. A ventilation pipe through an opening in the ceiling provided good ventilation (**Figure 4**). The temperature and pressures within each chamber could also be individually controlled [3].

In 1837, Pravaz built the largest hyperbaric chamber in Lyon, France, to seat 12 patients and treat patients with pulmonary conditions including tuberculosis, laryngitis, tracheitis, and pertussis, as well as unrelated conditions such as cholera,

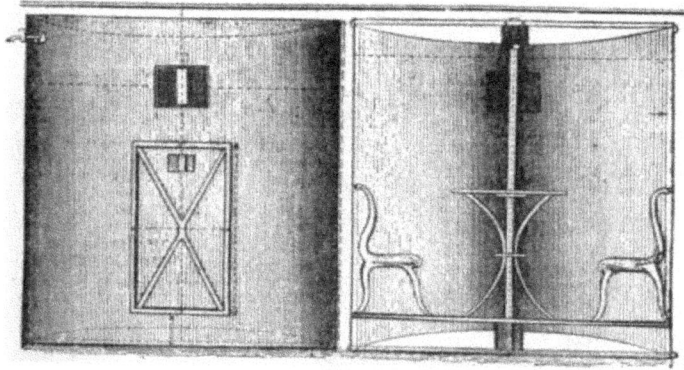

Figure 3.
Lange's pneumatic chamber.

Figure 4.
Leibig's pneumatic chamber.

conjunctivitis, deafness, menorrhagia, and rickets. In 1855, Bertin wrote a book on compressed air therapy and even constructed his own hyperbaric chamber.

Compressed air therapy was first introduced into the USA by JL Corning in 1871. In 1876, Kelly treated a patient in a "Compressed Air Bath Apparatus" having two locking plates operated from outside to seal pressures. In 1877, French surgeon Fontaine developed the first mobile hyperbaric operating theater. The high ambient pressure was claimed to facilitate the reduction of hernias and provide relief for patients with lung diseases. Over the next 3 months, 27 surgeries were successfully performed within this mobile hyperbaric chamber (**Figure 5**). Spurred by the results, Fontaine ventured to erect a mammoth hyperbaric surgical amphitheater to accommodate 300 patients in one sitting. This did not see the light of day as Fontaine died from an accident at the Pneumatic Institute to become the first physician to be martyred in the history of hyperbaric medicine [4].

In 1885, C Theodore Williams published his "Lectures on the Compressed Air Bath and its Uses in the Treatment of Disease" in the *British Medical Journal*, extolling the use of atmospheric air under different degrees of atmospheric pressure to treat diseases. He remarked that this mode of therapy was among the most important advances in modern medicine and expressed astonishment at its being ignored in England [5].

Back in the USA, during the closing days of the World War I, Kansas-bavsed physician Orval J Cunningham built a hyperbaric chamber in 1921 at Lawrence, Kansas. He used the facility to treat victims of the Spanish influenza epidemic that

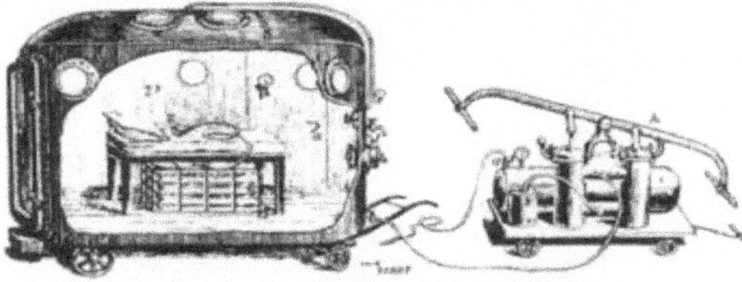

Figure 5.
Fontaine's mobile hyperbaric operation theater.

Figure 6.
Cunningham's hyperbaric hotel—outside and inside view.

swept North America. Noticing that people in the valley fared better than those living in the mountains, Cunningham theorized that atmospheric pressure or barometric factors were responsible for the higher mortality rates in those residing at higher elevations. He observed remarkable improvements in patients treated with HBO, especially those who were cyanotic and comatose. In 1923, heat from open gas burners warming the chambers in winter scorched the insulation and started a fire, but all patients were safely evacuated. In another incident, a mechanical failure caused a complete loss of pressure within the chamber and all patients died. This did not, however, deter Cunningham's enthusiasm for hyperbaric air. He went on to treat diseases such as syphilis, hypertension, diabetes mellitus, and cancer, believing that anaerobic infections played a role in the etiology of all these afflictions. In 1928, with the financial backing of Henry H. Timken, a roller bearing manufacturer and tycoon, Cunningham built the largest hyperbaric chamber in the world along the shores of Lake Erie in Cleveland, Ohio, at a cost of 1 million dollars. This "Steel Ball Hospital" or "Cunningham's Sanitarium" was a five-story high steel sphere, 64 ft in diameter with 60 rooms and weighing 900 tons. Each floor of this structure had 12 rooms, with all the amenities of a good hotel (**Figures 6** and 7). The growing popularity of Cunningham's treatments prompted the Bureau of Investigation of the American Medical Association (AMA) to request the doctor to validate his claims

Figure 7.
Cunningham's hyperbaric hotel—exterior and interior views.

regarding the effectiveness of hyperbaric therapy. Cunningham refused to share the details or cooperate with the AMA, leading to his being labeled a quack and a fraud. The chamber was dismantled in 1937 and sold for scrap during World War II [6].

3. A historical account of decompression sickness and its treatment

In 1840, Charles Pasley, charged with the recovery of the sunken warship HMS Royal George, commented that, of those who made frequent dives, "not a man escaped the repeated attacks of rheumatism and cold." In 1841, Trigger, a French mining engineer, used a pressure chamber to deliver workers to the bottom of the river to extract coal. In 1845, he reported that some of his miners complained of joint pains and nervous disorders after surfacing. The first recorded death from "caisson disease" (which later came to be known as decompression illness (DCI) or acute decompression sickness) occurred in 1859 during the building of the Royal Albert Bridge, a railway bridge in England spanning the River Tamar from Saltash to Plymouth. Several workers were taken ill after emerging from deep underground after long hours of work under high atmospheric pressure conditions. In 1871, during the construction of the Eads Bridge in St. Louis, 352 compressed air workers, including Alphonse Jaminet, the physician in charge, were employed. Thirty workers developed serious conditions with 12 ending fatally. Jaminet himself suffered decompression sickness, and his personal description was the first such recorded. It was in 1873 that Andrew Smith first utilized the term "caisson disease" to describe 110 cases of decompression sickness that occurred during construction of the Brooklyn Bridge. The project employed 600 compressed air workers, and recompression treatment was not available on site. In 1882, during the Hudson tunnel construction in New York, every fourth worker died of bends until a recompression chamber was installed to treat the condition. Only three workers died of bends over the next 18 months.

Paul Bert, a French professor of physiology and a student of Claude Bernard, is considered the father of pressure physiology (**Figure 8**). In 1878, while working closely with Dr. Alphonse Gal, the first doctor to actually dive in order to study

Figure 8.
Dr. Paul Bert (1883–1886).

how the body reacted underwater, Bert studied Gal's reports on divers who became symptomatic or died while surfacing. He conducted a series of dog experiments, exposing them to 7–9¾ atmospheres and subjecting them to rapid decompression. A majority of them died and exhibited grossly distended bodies with their right heart chambers filled with gas. When decompression was done at slowly over 1–2 hours after exposure to similar pressures, none of the dogs succumbed. Applying Dalton's and Henry's gas laws, Bert concluded that too rapid a decompression induced a pathophysiologic insult secondary to supersaturation of body tissues with nitrogen, causing the formation of nitrogen bubbles. He also went on to suggest that divers stop halfway to the surface to allow for slow decompression after a deep dive—what is now known as deep stops. Bert was also the first to describe oxygen toxicity at pressures in excess of 1.75 ATA. This adverse effect on the central nervous system came to be known as the "Paul Bert effect" [7, 8].

In 1908, Scottish physiologist John Scott Haldane conducted experiments at the Lister Institute of Preventive Medicine in London assisted by Lieutenant Guybon Damant of the Royal Navy, an expert diver and amateur scientist, and a physiologist Edwin Arthur Boycott. A herd of 85 goats was assembled, and the researchers put groups of up to eight goats inside compression chambers for specific periods of time. Pressures were then normalized before releasing the animals into the institute's yard for observation. These studies confirmed that those goats decompressed by stages did not exhibit signs of the bends (**Figure 9**). Haldane then introduced the concept of half times—the time required for a particular tissue to become half saturated with

Figure 9.
Bends in the foreleg of a goat after experiments performed by physiologist Haldane [9].

a gas—and recommended staged decompression, especially at shallower depths. He prepared detailed practical dive tables for the Royal Navy to prevent acute decompression sickness. These guidelines remained the foundation of all diving operations until 1956 [10]. Heinrich Drager was the first to explore the use of pressurized oxygen in decompression sickness (**Figure 10**). His protocols were put into practice by Behnke and Shaw, who used HBOT for treating decompression sickness in the late 1930s. They replaced oxygen in place of compressed air, and their work resulted in the use of the first nitrogen-oxygen mixtures and hyperbaric treatment being tailored to the severity of the injury [11]. In 1939, the US Navy began treating divers suffering decompression sickness with hyperbaric oxygen therapy. After World War II, the US military conducted extensive research in HBOT, and this expanded the existing knowledge about survivable pressures and popularized HBOT in the late 1950s and early 1960s. In the 1980s, Paul Harch began an in-depth study of brain decompression illness (DCI) and evaluated divers with this disorder. He concluded that it was not residual gas that was being treated but ischemic brain injury. He went on to develop individualized treatment protocols for over 50 different chronic neurological disorders. Harch is considered to be the foremost authority in the use of HBOT and SPECT brain blood flow imaging in neurology [12–14]. In 1990, former microbiology professor Igor Gamow invented and patented the Gamow Bag that provided mountaineers with a mobile and effective method to treat high-altitude sickness. This bag is a single-place portable hyperbaric chamber, pressurized with a

Figure 10.
Drager and his recompression chamber.

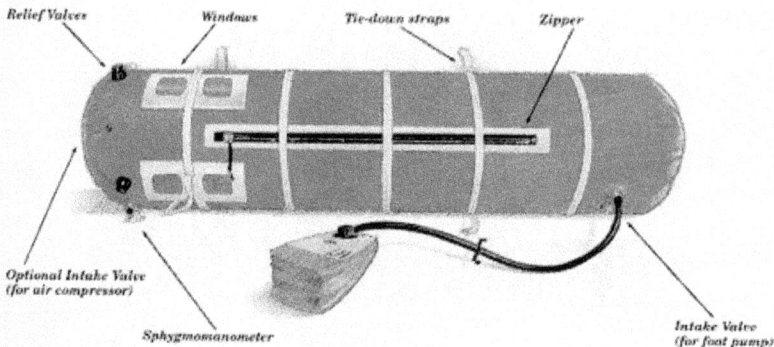

Figure 11.
The Gamow Bag.

foot pump, to simulate a descent to 7000 ft (**Figure 11**). In 1992, Harch treated the first delayed decompression sickness, which led to the treating of "dementia pugilistica" in boxers and cerebral palsy and autism in children [15].

4. Treating diseases with HBO

In 1937, Brazilians Ozorio de Almeida and Costa pioneered the use of HBOT in treating leprosy [16]. In the 1950s, Ite Boerema, a cardiac surgeon from the Netherlands, conceived the idea of "flooding" the body's tissues with extra oxygen. Working with the help of the Royal Dutch Navy, Boerema conducted a series of animal experiments and operations within a hyperbaric oxygen chamber (**Figure 12**). These went off without a hitch and led to the installation of a large operating hyperbaric chamber at the University of Amsterdam. Many children with congenital heart diseases like tetralogy of Fallot, transposition of great vessels, and pulmonic stenosis were operated in this facility with great success. Boerema mooted the concept of "Life without blood" using HBO, when dissolved oxygen sufficed to meet the entire body's oxygen needs without the need for red cells or hemoglobin. Boerema is credited with being the father of modern-day hyperbaric medicine [17].

In 1955–1956, I Churchill-Davidson evaluated clinical trials on HBOT as a potentiator for radiation therapy in cancer patients at St. Thomas Hospital in London [18]. Public interest in hyperbaric oxygen therapy started to grow in the 1960s after publicity about its use in treating President John F Kennedy's sick infant. In 1961, a colleague of Boerema, W. H. Brummelkamp, published a paper on inhibition of anaerobic infections by HBOT [19]. In 1962, Smith and Sharp reported the enormous benefits of HBO in carbon monoxide poisoning. They recommended that all those having a verified carboxyhemoglobin level above 25% needed immediate HBOT at 3 ATA for 90 minutes, followed by two or three more sessions for full recovery, making HBO very cost-effective [20]. Global interest in HBOT was rekindled by this finding, resulting in hyperbaric units being installed at many centers like Duke University, New York Mount Sinai Hospital, Presbyterian Hospital and Edgeworth Hospital in Chicago, Good Samaritan in Los Angeles, St. Barnabas Hospital in New Jersey, Harvard Children's Hospital, and St. Luke's Hospital in Milwaukee. In 1965, Perrins from the UK demonstrated the effectiveness of HBOT

Figure 12.
Dr. Boerema with children operated by him.

in osteomyelitis [21]. In 1966, Saltzman and coworkers from the USA proved the effectiveness of HBOT in stroke patients [22].

In 1970, Boschetty and Cernoch of Czechoslovakia conducted a trial of HBOT for multiple sclerosis. In their series 15 out of 26 patients with multiple sclerosis showed improvement after HBOT at 2 atmospheres [23]. In 1971, Lamm of West Germany used HBOT for treatment of sudden deafness. It was shown that HBOT shortens the course of healing in high-pitch perception dysacusis by upregulating constitutive nitric oxide synthase in the substructure of the cochlea [24]. In 1973, Thurston pioneered studies that showed lower mortality figures in patients with myocardial infarction treated with HBO. HBOT was shown to improve oxygen supply to the threatened heart and reduce the volume of infarct size and other major adverse outcomes [25]. In 1972, Richard A Neubauer set up the Ocean Hyperbaric Neurologic Center in Lauderdale-by-the-Sea exclusively for HBOT in the management of various central nervous system disorders. He mooted the concept of "idling" neurons capable of surviving for years or even decades after the original injury. He claimed that these injured neurons could be re-activated with HBOT and that the greater the number of idling neurons, the better would be the patient's response to HBOT [26]. Neubauer was also the co-founder and executive director of the American College of Hyperbaric Medicine. After his death in 2007 at the age of 83, his clinical research center in Florida was renamed the Neubauer Hyperbaric Neurologic Center. In 1976, Hollbach and Wasserman determined that 1.5 ATA (atmospheres absolute) maximizes oxygen content and glucose metabolism in the brain [27].

In 1985, RE Marx and his colleagues observed that the rate of osteoradionecrosis was 30%/patient in patients treated with penicillin alone while rates in those treated with HBO was only 5% [28]. In 1987, Jain successfully treated patients with paralytic stroke using HBOT [29, 30]. In 2002, a US Army study confirmed that HBOT repairs white matter damage in children with cerebral palsy. In 2005, Stoller of the USA treated the first case of a child with fetal alcohol syndrome using HBOT and with good outcome [31]. In 2006, Thom of the USA discovered that HBO causes stem cell mobilization [32]. In 2010, Godman discovered that HBOT activated 8101 genes, resulting in reduction of inflammation and increase in growth in body tissues [33, 34]. In 2011, Stoller treated the first retired National Football League (NFL) player for chronic traumatic encephalopathy [35]. In 2012, Harch and his colleagues demonstrated that blast-induced post-concussion syndrome and post-traumatic stress disorders responded to HBOT [15].

Air or gas embolism
Carbon monoxide poisoning; cyanide poisoning; smoke inhalation
Clostridial myositis and myonecrosis (gas gangrene)
Crush injuries, compartment syndromes, and other acute traumatic peripheral ischemias
Decompression sickness
Enhancement of healing in selected problem wounds
Exceptional blood loss anemia
Intracranial abscess
Necrotizing soft tissue infections
Refractory osteomyelitis
Skin flaps and grafts (compromised)
Delayed radiation injury (soft tissue and bony necrosis)
Thermal burns

Table 1.
UHMS- and FDA-approved indications for hyperbaric oxygen therapy.

The UHMS and FDA approved HBOT for treatment of conditions like autism, stroke, air embolism, ischemic limbs, split-thickness skin graft acceptance, failed grafts, flap survival and salvage, wound reepithelialization, acute thermal burns, etc. (**Table 1**) [36, 37].

Many patients do not respond to aggressive acid-suppressing medications. HBOT has a beneficial effect in patients with blunt duodenal trauma, duodenal ulcers, and indomethacin-/radiation-induced gastritis. This salubrious effect is mediated by decreased production of oxidative stress markers like tumor necrosis factor-alpha, interleukin-1beta, neopterin, myeloperoxidase, and malondialdehyde. HBOT is seen to improve the acid-neutralizing function of the stomach, normalize gastric motility, reduce the duodenum acidification, decrease edema, and improve the blood flow both in human and equine studies [38, 39]. These effects were also seen in cases of inflammatory bowel diseases like Crohn's [40].

5. Landmark academic events in HBOT

In September 1961, the First International Congress on the clinical applications of hyperbaric oxygen was held in Amsterdam. The Second International Conference on HBO was held in Glasgow in September 1964, with detailed deliberations on various aspects of HBOT. In November 1965, the Third International Congress on HBOT was organized at the Duke University at Durham, North Carolina. The Fourth and Fifth International Congresses were held in Sapporo, Japan, and Vancouver, respectively, in 1969 and 1973. The University of Aberdeen, Scotland, hosted the sixth conference in August 1977. The subsequent International Congress was held in Moscow in 1981 and is an annual event thereafter. The deliberations during these academic forums threw fresh light on the basic physiology, oxygen toxicity, and therapeutic applications of HBO in human disease.

The Undersea Medical Society (it added hyperbaric to its name in 1986), an organization made up largely of naval and ex-navy physicians, was founded in 1967 in the USA. It reviewed the indiscriminate and inappropriate use of the HBO chamber for a variety of medical conditions by practitioners searching for a "cure-all" therapy, tarnishing the credibility of hyperbaric medicine. This non-profit organization, now known as the Undersea and Hyperbaric Medical Society (UHMS), set up a Committee on Hyperbaric Oxygen Therapy in the 1970s to systematically review all the available scientific evidence for HBOT and formulate absolute indications for HBOT. This was accepted by insurance providers, including Medicare. The UHMS is committed to providing, promoting, developing, and raising the quality of care across the spectrum in scientific communication, life sciences, and clinical practices of hyperbaric medicine by promoting high standards of patient care and operational safety. It offers accreditation and certificate of competency and credibility and has over 2500 members in 50 countries. UHMS also awards board certification in Undersea and Hyperbaric Medicine through the American Board of Emergency Medicine (ABEM), the American Board of Preventive Medicine (ABPM), and fellowship training in Undersea and Hyperbaric Medicine.

In 1980, Dr. Richard A. Neubauer and Dr. William S. Maxfield formed the American College of Hyperbaric Medicine (ACHM) to foster the ethical advancement and expansion of hyperbaric medicine. The International Society of Hyperbaric Medicine was founded in 1988.

Hyperbaric medicine was approved by the American Board of Medical Specialties as a sub-specialty of emergency and preventative medicine in 2000.

6. Developments in HBOT chambers

In 1860, the first hyperbaric chamber in the North American continent was constructed in Oshawa, Ontario, Canada. A year later, a neurologist, James Leonard Corning, built the first hyperbaric chamber in the USA in New York. This chamber was used to treat "nervous and related disorders." The first decompression chamber was invented by the Italian engineer Alberto Gianni in 1916 [39, 40]. In 1928, the Harvard Medical School built a hyperbaric chamber for medical research. Among the largest HBOT chambers is the 22 ton 32 ft wide 14 ft wide one at the Utah Valley Regional Medical Center, USA.

In modern times, many traditional hard-shell hyperbaric chambers and soft-shell, portable hyperbaric chambers (**Figures 13–18**) are manufactured by several companies and available in every major city. The latest chambers must comply with NFPA-992012 Edition Chapter 14 Code in the USA and European 1997 CEN pressure vessels 97/23E standards as well as the 1998 ECHM recommendations for safety. The newer chambers feature hingeless pressure-sealed doors, antifriction bearings, antibacterial leather upholstery, and high-quality resin fiber loop mattresses and pillows that dissipate heat and moisture generated by the body during therapy. The newer low-pressure monoplace chambers are portable and less

Figure 13.
The evolution of hyperbaric chambers.

Figure 14.
Monoplace HBO chamber.

Figure 15.
Recompression chamber.

Figure 16.
Multiplace HBOT chamber.

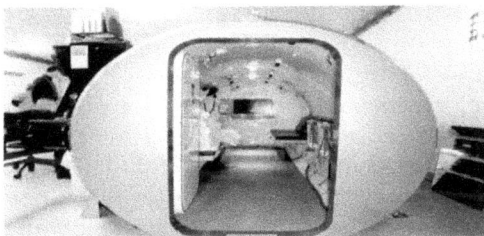

Figure 17.
Hyperbaric operation suite.

expensive. Operating between 1.2 and 1.3 ATA pressures, they are eminently suited for use in homes and spas and also find use to improve results after plastic surgery.

The earliest documentation of therapeutic use of HBOT in animals was in 1998. The Veterinary Hyperbaric Medicine Society was formed in 2006. Veterinary-specific hyperbaric chambers are available.

The evolution of HBOT chambers over time is chronicled in **Figure 13**.

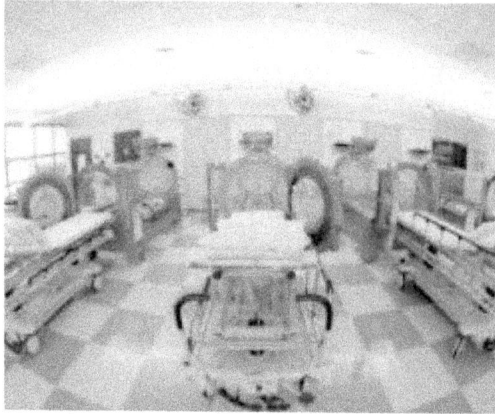

Figure 18.
Hyperbaric operation suite. (Image courtesy: CONE Health Wound Care and Hyperbaric Center, Greensboro, North Carolina, USA).

7. Current status of HBOT

HBOT was called the Cinderella of modern medicine since it was not taught in medical schools and had no pharmaceutical companies to nurture and protect it. Over the course of time, it has shed the label of being a mysterious therapy and become a major tool in the armamentarium of clinicians either as a primary or adjunctive therapy for a spectrum of diseases. Stroke, cancer, heart disease, and chronic lung disease account for almost 60% of the total number of deaths. Hypoxia is a significant component of the pathology of these conditions, and this leads to metabolic acidosis, organ dysfunction, and death. Conventional oxygen therapy may not have desired results, when HBOT yields remarkable clinical improvement. HBOT prevents 75 percent of all major amputations that would otherwise be necessary for diabetic wounds and a 450% increase in complete recovery in patients with traumatic brain injury receiving HBOT vs. standard intensive care. Newer application of HBOT is in emergency care for resuscitation in cases of acute blood loss, near drowning, hanging and poisoning, and cardiorespiratory arrest.

Athletic associations like the NFL employ hyperbaric oxygen therapy as part of the recovery regimen for its athletes, and some players own their own HBOT chambers. Joe Namath experienced remarkable recovery from the head injuries he sustained during his career, leading him to be part of an FDA-approved study of HBOT at the Joe Namath Neurological Center of the Jupiter Medical Center in Florida. Ace swimmer Michael Phelps and football stars Maurice Jones-Drew and James Harrison have endorsed the benefits of HBOT, along with professional boxers like Evander Holyfield [41].

8. Conclusions

With the utilization of isotopic tracers, magnetic resonance imaging (MRI), and single-photon emission computed tomography (SPECT), HBOT is getting evidence-based recognition. Various conditions like brain injuries, stroke, and neurological diseases with poor prognosis are now amenable to improved outcomes with the application of HBOT. There are more than 500 hyperbaric facilities in the USA alone. Much research remains to be done regarding the efficacy of HBO_2 therapy to develop treatment plans for those in extremes of age. The use of hyperbaric medicine to treat wounds in the foot or in the brain is a divine gift, and great advances in this field are on the horizon. The future of healthcare is here!

Author details

Chandrasekhar Krishnamurti
Department of Anesthesiology, NRI Institute of Medical Sciences, Visakhapatnam, Andhra Pradesh, India

*Address all correspondence to: globeshaker@gmail.com

IntechOpen

References

[1] History of Hyperbaric Oxygen Therapy, Richmond Hyperbaric Health Centre. www.richmond-hyperbaric.com

[2] Jain KK. Textbook of Hyperbaric Medicine. 4th ed. Hogrefe & Huber: Ashland, OH; 2004

[3] Tissier PLA. In: Cohen SS, editor. Pneumotherapy: Including Aerotherapy and Inhalation Methods. Vol. X. Philadelphia: P. Blakiston's Sons and Co; 1903

[4] Stewart J Jr, Corning JL. Exploring the History of Hyperbaric Chambers, Atmospheric Diving Suits and Manned Submersibles: The Scientists and Machinery. Bloomington, IN: Xlibris; 2011. p. 68. Available from: http://bit.ly/1o6LZhU [Accessed: 14 May 2014]

[5] Theodore Williams C. Lectures on the compressed air bath and its uses in the treatment of disease. British Medical Journal. 1885;**1268**(1):769-772

[6] Choffin M. The Cunningham Sanitarium, Cleveland Historical. Available from: https://clevelandhistorical.org/items/show/378 [Accessed: 31 January 2019]

[7] History of Hyperbaric Medicine and HBOT. hbot.g7oz.org

[8] Bert P. La Pression Barométrique, Recherches de Physiologie Expérimentale (1878), Barometric Pressure: Researches in Experimental Physiology; 1943

[9] Haldane JS. Journal of Hygiene: Cambridge University Press. 1908;**8**(2)

[10] Sekhar KC, Chakra Rao SSC, Haldane JS. The father of oxygen therapy. Indian Journal of Anaesthesia. 2014;**58**(3):350-352

[11] Clark D. History of hyperbaric therapy. In: Neuman TS, Thom SR, editors. Physiology and Medicine of Hyperbaric Oxygen Therapy. 1st ed. Philadelphia, PA: Saunders; 2008. pp. 3-18

[12] Harch PG, Neubauer RA. Hyperbaric oxygen therapy in global cerebral ischemia/anoxia and coma. In: Jain KK, editor. Chapter 19, Textbook of Hyperbaric Medicine. 5th Revised ed. Seattle, WA: Hogrefe and Huber Publishers; 2009. pp. 235-274

[13] Harch PG, Neubauer RA. Hyperbaric oxygen therapy in global cerebral ischemia/ anoxia and coma. In: Jain KK, editor. Chapter 18, Textbook of Hyperbaric Medicine. 3rd Revised ed. Seattle, WA: Hogrefe and Huber Publishers; 2004. pp. 223-261

[14] Harch PG, Neubauer RA, Uszler JM, James PB. Appendix: Diagnostic imaging and HBO therapy. In: Jain KK, editor. Chapter 44, Textbook of Hyperbaric Medicine. 5th Revised ed. Seattle, WA: Hogrefe and Huber Publishers; 2009. pp. 505-519

[15] Harch PG, Fogarty EF, Staab PF, van Meter K. Low pressure hyperbaric oxygen therapy and SPECT brain imaging in the treatment of blast-induced chronic traumatic brain injury (post-concussion syndrome) and post traumatic stress disorder: A case report. Cases Journal. 2009;**2**:6538

[16] De Alemeida AO, Rabello E. Dermatological Society, National Academy of Medicine of Rio de Janeiro; 1937

[17] Leopardi LN, Metcalfe MS, Fortde A, et al. Ite Boerema—Surgeon and engineer with a double-Dutch legacy to medical technology. Surgery. 2004;**135**(1):99-103

[18] Neubauer RA, Maxfield WS. The polemics of hyperbaric medicine. Journal of the American Physicians and Surgeons. 2005;**10**:15-17

[19] Brummelkamp WH, Hogendijk L, Boerema I. Treatment of anaerobic infections (Clostridial myositis) by drenching the tissues with oxygen under high atmospheric pressure. Surgery. 1961;**49**:299-302

[20] Smith G, Sharp GR. Treatment of carbon-monoxide poisoning with oxygen under pressure. Lancet. 1960;**276**:905-906

[21] Perrins JD, Maudsley RH, Colwill MW, Slack WK. Thomas DAOHP in the management of chronic osteomyelitis. In: Brown IW, Cox BG, editors. Proceedings of the Third International Conference on Hyperbaric Medicine. Washington, DC: National Academy of Sciences, National Research Council; 1966. pp. 578-584

[22] Saltzman HA, Heyman A, Whalen RE. The use of hyperbaric oxygen in the treatment of cerebral ischemia and infarction. Circulation. 1966; **33 & 34**(Supplement II):II-20-II-27

[23] Boschetty V, Cernoch J. Aplikace kysliku za pretlaku a nekterych neurologickych onemocneni. Vol. 53. 1980. pp. 687-690

[24] Lamm K, Lamm H, Arnold W. Effect of hyperbaric oxygen therapy in comparison to conventional or placebo therapy or no treatment in idiopathic sudden hearing loss, acoustic trauma, noise-induced hearing loss and tinnitus. A literature survey. Advances in Oto-Rhino-Laryngology. 1998;**54**:86-99

[25] Thurston GJ, Greenwood TW, Bending MR, Connor H, Curwen MP. A controlled investigation into the effects of hyperbaric oxygen on mortality following acute myocardial infarction. Quarterly Journal of Medicine. 1973;**XLII**:751-770

[26] Neubauer RA, Gottlieb SF, Kagan RL. Enhancing "idling" neurons. Lancet. 1990;**335**:542

[27] Holbach KH, Caroli A, Wassmann H. Cerebral energy metabolism in patients with brain lesions of normo- and hyperbaric oxygen pressures. Journal of Neurology. 1977;**217**(1):17-30

[28] Marx RE, Johnson RP, Kline SN. Prevention of osteoradionecrosis: A randomized prospective clinical trial of hyperbaric oxygen versus penicillin. Journal of the American Dental Association. 1985;**111**(1):49-54

[29] Jain KK. Hyperbaric oxygen in acute ischemic stroke. Stroke. 2003;**34**(2):571-574

[30] Jain KK. Role of hyperbaric oxygenation in the management of stroke. In: Jain KK, editor. Textbook of Hyperbaric Medicine. 3rd ed. Göttingen, Germany: Hogrefe & Huber; 2003

[31] Stoller KP. Quantification of neurocognitive changes before, during, and after hyperbaric oxygen therapy in a case of fetal alcohol syndrome. Pediatrics. 2005;**116**:e586-e591

[32] Thom SR, Bhopale VM, Velazquez OC, et al. Stem cell mobilization by hyperbaric oxygen. American Journal of Physiology. Heart and Circulatory Physiology. 2006;**290**:H1378-H1386

[33] Godman C, Chheda K, Hightower L, Perdrizet G, Shin D-G, Giardina C. Hyperbaric oxygen induces a cytoprotective and angiogenic response in human microvascular endothelial cells. Cell Stress & Chaperones. 2010;**15**:431-442

[34] Godman CA, Joshi R, Giardina C, Perdrizet G, Hightower LE. Hyperbaric oxygen treatment induces antioxidant gene expression. Annals of the New York Academy of Sciences. 2010;**1197**:178-183

[35] Stoller KP. Hyperbaric oxygen therapy (1.5 ATA) in treating sports

related TBI/CTE: Two case reports. Medical Gas Research. 2011;**1**:17

[36] Hampson NB, editor. Hyperbaric Oxygen Therapy: 1999 Committee Report. Kensington, MD: Undersea and Hyperbaric Medical Society; 1999

[37] Gesell LB, editor. Hyperbaric Oxygen Therapy Indications. The Hyperbaric Oxygen Therapy Committee Report. 12th ed. Durham, NC: Undersea and Hyperbaric Medical Society; 2008

[38] Güneş AE, Gözeneli O, Akal A, Taşkın A, Sezen H, Güldür ME. An experimental study on the effectiveness of hyperbaric oxygen and thymoquinone treatment in blunt duodenal injury. Konuralp Tıp Dergisi. 2018;**10**(3):347-353

[39] Yang Z, Nandi J, Wang J, Bosco G, et al. Hyperbaric oxygenation ameliorates indomethacin-induced enteropathy in rats by modulating TNF-α and IL-1β production. Digestive Diseases and Sciences. 2006;**51**(8):1427-1433

[40] Rossignol DA. Hyperbaric oxygen treatment for inflammatory bowel disease: A systematic review and analysis. Medical Gas Research. 2012;**2**:6

[41] Kindwall EP. A history of hyperbaric medicine. In: Hyperbaric Medicine Practice. 3rd ed. Flagstaff, AZ: Best Publishing; 2008. pp. 3-22